I0053871

Engineering Education and
Technological / Professional Learning

Engineering Education and Technological / Professional Learning

Special Issue Editors

Clara Viegas
Arcelina Marques
Gustavo R. Alves
Francisco José García-Peñalvo

MDPI • Basel • Beijing • Wuhan • Barcelona • Belgrade

MDPI

Special Issue Editors

Clara Viegas
Instituto Superior de
Engenharia do Porto
Portugal

Arcelina Marques
Instituto Superior de
Engenharia do Porto
Portugal

Gustavo R. Alves
Instituto Superior de
Engenharia do Porto
Portugal

Francisco José García-Peñalvo
University of Salamanca
Spain

Editorial Office
MDPI
St. Alban-Anlage 66
4052 Basel, Switzerland

This is a reprint of articles from the Special Issue published online in the open access journal *Education Sciences* (ISSN 2227-7102) from 2018 to 2019 (available at: https://www.mdpi.com/journal/education/special_issues/Engineering_Education_Technological).

For citation purposes, cite each article independently as indicated on the article page online and as indicated below:

LastName, A.A.; LastName, B.B.; LastName, C.C. Article Title. *Journal Name* **Year**, *Article Number, Page Range.*

ISBN 978-3-03921-984-1 (Pbk)
ISBN 978-3-03921-985-8 (PDF)

© 2019 by the authors. Articles in this book are Open Access and distributed under the Creative Commons Attribution (CC BY) license, which allows users to download, copy and build upon published articles, as long as the author and publisher are properly credited, which ensures maximum dissemination and a wider impact of our publications.

The book as a whole is distributed by MDPI under the terms and conditions of the Creative Commons license CC BY-NC-ND.

Contents

About the Special Issue Editors

Clara Viegas (PhD) graduated in Physics/Applied Mathematics and has an MSc degree in Mechanical Engineering from the University of Porto. She holds a PhD in Science and Technology (Physics Didactics) from the University of Trás-os-Montes e Alto Douro. She has been a professor at the Polytechnic of Porto School of Engineering, Portugal, since 1994. She was a member of the Physics Department Management Board between 2008 and 2012, and was Vice-President between 2002 and 2005. She served as a member of the Pedagogical Council between 2007 and 2009. She is a researcher at the CIETI-LABORIS Centre for Innovation in Engineering and Industrial Technology, and a participant in several national and international R&D projects in areas of science and engineering education. An invited chair in international conferences and referee in JCR journals and an author of more than 80 papers in international scientific journals, books, and international conferences with peer review, she co-authored a book, co-edited two books, and a special issue in an international journal. Her research interests include engineering education, physics didactics, teacher mediation, professional development, and remote experimentation learning outcomes.

Arcelina Marques (PhD) graduated in Physics from the University of Porto, and has an MSc degree in Physics of Laser Communications from the University of Essex from 1992 and a PhD degree in Engineering Sciences from the University of Porto from 2008. She has been with the Polytechnic of Porto School of Engineering (ISEP) since 1995, where she lectures physics, electronics, and biomechanics courses. She has been involved in several R&D projects in the areas of biomechanics, medical instrumentation, engineering education, and remote experimentation, which are her present research interests. She is the author and co-author of more than 60 conference and journal papers, one national patent, and one book chapter. She served as the Programme Chair of the 11th Remote Engineering and Virtual Instrumentation (REV2014) conference, and has been a member of several program scientific committees of international conferences in the above research fields. She was a member of the Physics Department Management Board from 2008 to 2012 and 2016 to 2018. At present, she serves as the vice-president of ISEP' Scientific Council.

Gustavo R. Alves (PhD) graduated in 1991 and obtained an MSc and a PhD degree in Computers and Electrical Engineering in 1995 and 1999, respectively, from the University of Porto, Portugal. He has been a professor at the Polytechnic of Porto School of Engineering since 1994. He has authored or co-authored over 250 publications, including book chapters and conference and journal papers with a referee process, and also co-edited two books. He was involved in 19 national and international research projects. His research interests include engineering education, remote experimentation, and designs for debugging and testing. He served as program co-chair of the 1st and 2nd International Conferences of the Portuguese Society for Engineering Education (CISPEE2013 and CISPEE2016), of the 3rd Experiment@ International Conference, and as general chair of the 11th Remote Engineering and Virtual Instrumentation (REV2014) conference, and of the 3rd Technological Ecosystems for Enhancing Multiculturality (TEEM2015) conference. He is currently serving as general co-chair of EDUCON2020, TAEE2020, and CISPEE2020. Dr. Alves currently serves as an Associated Editor for the IEEE Transactions on Learning Technologies and the IEEE Journal of Latin-American Learning Technologies (IEEE-RITA).

Francisco José García-Peñalvo (PhD) received his bachelor's degree in computing from the University of Salamanca, Spain, and the University of Valladolid, Spain, and his Ph.D. degree from the University of Salamanca. He was the Vice Chancellor of technological innovation with the University of Salamanca from 2007 to 2009. He is currently the Director of the Research Group in interaction and e-Learning at the University of Salamanca. He has led and participated in over 50 research and innovation projects. He has published over 200 articles in international journals and conferences. His main research interests include e-Learning, computers and education, adaptive systems, web engineering, semantic web, and software reuse. He coordinates the Doctoral Program in Education at the Society of Knowledge at the University of Salamanca. He has been a Guest Editor of several special issues of international journals, including Online Information Review, Computers in Human Behavior, and Interactive Learning Environments. He is the Editor-in-Chief of the Education in the Knowledge Society magazine and the Journal of Information Technology Research.

education sciences

MDPI

Editorial

Engineering Education and Technological/Professional Learning

Clara Viegas [1,*], Arcelina Marques [1], Gustavo R. Alves [2] and Francisco García-Peñalvo [3]

[1] Research Centre in Industrial Technology and Engineering/CIETI-Laboris/Physics Department, ISEP, School of Engineering/Polytechnic of Porto, Rua Dr. António Bernardino de Almeida, 431, 4249-015 Porto, Portugal; mmr@isep.ipp.pt

[2] Research Centre in Industrial Technology and Engineering/CIETI-Laboris/Electrical Engineering Department, ISEP, School of Engineering/Polytechnic of Porto, Rua Dr. António Bernardino de Almeida, 431, 4249-015 Porto, Portugal; gca@isep.ipp.pt

[3] Research Institute of Educational Sciences/IUCE, Computer Science Department, University of Salamanca/Paseo de Canalejas, 169, Edificio Solís, 1ªPlanta, 37008 Salamanca, Spain; garcia@usal.es

* Correspondence: mcm@isep.ipp.pt; Tel.: +351-228-040-500

Received: 18 November 2019; Accepted: 18 November 2019; Published: 22 November 2019

check for updates

The focus of this Special Issue is aimed at enhancing the discussion of Engineering Education, particularly related to technological and professional learning. It was associated with TEEM'18 (6th International Conference on Technological Ecosystems for Enhancing Multiculturality), in particular the special track (under the same nomination) [1] from which the authors of the best works were invited to submit an extension of their paper. Later, it was also associated with CASHE'19 (1st Conference on Academic Success in Higher Education). In parallel, since this special issue had an open call, several works were submitted, from different universities and countries (Australia, Canada, Japan, Portugal, Spain, United Arab Emirates, United Kingdom), creating a more global view of different, yet similar concerns.

A special thanks to all authors who responded to this call, contributing to the compiled multicultural vision on this theme. The quality of the works received was carefully scrutinized by a panel of international reviewers. To all of them we would like to express our sincere appreciation. Even though not all of the received works were accepted (from 18 submissions, we selected 10 papers to include), the quality of the papers received attests to the significance of debating this theme.

Here the reader can find works tackling several interesting topics such as: Educational resources addressing students' development of competences, the importance of final year projects as a link to the professional environments, professional project management competences, the importance of multicultural and interdisciplinary challenges, sustainable product design focusing on future professional menaces, and ways of improving didactical issues aimed at students' involvement and their development as future engineers.

In the 21st century, students face a challenging demand: They are expected to have the best scientific expertise, but also highly-developed social skills and qualities like teamwork, creativity, communication and leadership. Even though students should be prepared through their academic education, there is still a gap between academic life and professional life. This gap is usually fulfilled with informal learning provided by older colleagues while these young engineers are immersed in the professional field. Though unavoidable, this gap can be lessened if students are already aware of some important working and social skills [2–4].

Engineering education organizations have been addressing new professional challenges, guided by general concerns, such as teamwork abilities, argumentation and persuasion abilities in multiple social contexts, creativity, complexity handling, leadership, and strong work ethics [2]. This stresses the importance of these competences being worked through college along with communication, scientific/technological expertise, problem-solving or analytical/quantitative skills. Nowadays, it is as important to address scientific expertise as students' social and professional competences. Even though both are important, they do not play equal parts in the minds of students and teachers [5–8]. Plus, the outcome perspectives from academic and professional worlds can be quite different. Thus, different points of view must be acknowledged and documented.

The need for a better understanding of engineering education in the 21st century is reflected in scientific research [4,5,7,8], where it is common to encounter big experiences, involving funding and school commitments. However, it is less common to encounter similar important scientific studies that can be applied by any good willing teacher. Smaller scale studies, representing better-contextualized teaching closer to professional demands, can also bring valid insight to this discussion in the scientific community. The purpose of this Special Issue was to help identify good practices and/or particular concerns that young engineers, their teachers or their employees feel needed improvement.

"Tutorials for Integrating CAD/CAM in Engineering Curricula" [9] talks about the importance of developing computer-based competences within engineering courses. The authors discuss how the use of specific tutorials helped their students solving engineering problems in real-life settings.

"Development of Final Projects in Engineering Degrees around an Industry 4.0-Oriented Flexible Manufacturing System: Preliminary Outcomes and Some Initial Considerations" [10] is also dedicated to digital engineering competences. It explains how an educational tool, made to develop final projects in an engineering degree, helped students exploring different aspects in parallel, such as automation, supervision, instrumentation, communication and robotics.

"Lessons Learned from the Development of Open Educational Resources at Post-Secondary Level in the Field of Environmental Modelling: An Exploratory Study" [11] explores a different perspective of our digital era: it claims students tend to seek more resources when they are enrolled in a course with online materials. Authors claim that if these open educational sources are well designed, students' achievements, involvement and satisfaction are very positive.

Another important 21st century engineering competence is project management. "Project Management Competences by Teaching and Research Staff for the Sustained Success of Engineering Education" [12] brings us a study about the importance of improving professional project management competences. Authors describe how research may be used to improve teachers' skills while guiding students into this process.

Professional and socio-professional engineering competences are addressed in "Fostering Professional Competencies in Engineering Undergraduates with EPS@ISEP" [13], under the scope of the European Project Semester. During a semester, students from different countries, degrees and cultures, come together and use their diversity of experiences and expertise to solve a problem through a Project Based Learning (PBL) experience.

Future professionals should also develop vital competences regarding rethinking products and business models in order to address the emergent sustainability problems. As the author of "Eco-Design and Eco-Efficiency Competencies Development in Engineering and Design Students" [14] explains, students may gain sensibility as they develop professional competences while working in assignments provided by real industrial companies with this concern in mind.

"Enhancing Railway Engineering Student Engagement Using Interactive Technology Embedded with Infotainment" [15] starts with a brief review on different teaching pedagogies addressing their suitability and use of interactive technologies. The authors discuss a new teaching application which they think can improve students' participation and performance.

Students' perceptions regarding their development through their education as future engineers are addressed in "Students' Perceptions Regarding Assessment Changes in a Fluid Mechanics Course" [16]. The authors chose a crucial aspect to analyze: the impact of different methods of students' assessment. Social and scientific competences are analyzed in parallel.

Educ. Sci. **2019**, *9*, 277

The importance of the learning objectives being completely understood by the students from the beginning of each course is the theme of "An Extended Constructive Alignment Model in Teaching Electromagnetism to Engineering Undergraduates" [17]. This work addresses the importance of a modification in students' perceptions of what each course may represent and how it can aid them as future professionals while developing strong consolidated competences throughout their degree.

The final year project is also addressed in "A Systematic Review of Project Allocation Methods in Undergraduate Transnational Engineering Education" [18]. In this work, several strategies of project allocations are identified studying the corresponding students' experiences and learning gains. Authors discuss how different factors can affect these allocations and they make recommendations in order to solve some of the identified challenges.

We believe these papers may provide an insightful reflection of our own practices as engineering educators. Hopefully, those reflections on particular aspects of our teaching may contribute to enhance students' development towards the upcoming challenges of their future career! In the meantime, work continues [19,20].

Conflicts of Interest: The authors declare no conflict of interest.

References

1. Viegas, C.; Marques, A.; Alves, G. Engineering Education and Technological/Professional Learning. In Proceedings of the 6th Technological Ecosystems for Enhancing Multiculturality (TEEM'18), Salamanca, Spain, 24–26 October 2018; ACM: New York, NY, USA, 2018; pp. 58–60. [CrossRef]
2. Johansen, B. *Leaders Make the Future: Ten New Leadership Skills for an Uncertain World*; Berrett-Koehler Publishers Inc.: San Francisco, CA, USA, 2012; ISBN 9781609944872.
3. Wood, R.; McGlashan, A.; Moon, C.-B.; Kim, W. Engineering Education in an Integrated Setting. *IJEP* **2018**, *8*, 17–27. [CrossRef]
4. Crawley, E.; Malmqvist, J.; Ostlund, S.; Ostlund, S.; Brodeur, D.R.; Edstrom, K. *Rethinking Engineering Education—The CDIO Approach*; Springer: New York, NY, USA, 2014. [CrossRef]
5. Malmi, L.; Adawi, T.; Curmi, R.; Graaff, E.; Duffy, G.; Kautz, C.; Kinnunen, P.; Williams, B. How Authors Did It—A Methodological Analysis of Recent Engineering Education Research Papers in the European Journal of Engineering Education. *Eur. J. Eng. Educ.* **2018**, *43*, 171–189. [CrossRef]
6. Borrego, M.; Bernhard, J. The emergence of engineering education research as an internationally connected field of inquiry. *J. Eng. Educ.* **2011**, *100*, 14–47. [CrossRef]
7. Huang-Saad, A.Y.; Morton, C.S.; Libarkin, J.C. Entrepreneurship Assessment in Higher Education: A Research Review for Engineering Education Researchers. *J. Eng. Educ.* **2018**, *107*, 263–290. [CrossRef]
8. Bernhard, J. Engineering Education Research in Europe—Coming of Age. *Eur. J. Eng. Educ.* **2018**, *43*, 167–170. [CrossRef]
9. Ullah, A.M.M.; Harib, K. Tutorials for Integrating CAD/CAM in Engineering Curricula. *Educ. Sci.* **2018**, *8*, 151. [CrossRef]
10. González, I.; Calderón, A. Development of Final Projects in Engineering Degrees around an Industry 4.0-Oriented Flexible Manufacturing System: Preliminary Outcomes and Some Initial Considerations. *Educ. Sci.* **2018**, *8*, 214. [CrossRef]
11. Hassan, Q.K.; Rahaman, K.R.; Sumon, K.Z.; Dewan, A. Lessons Learned from the Development of Open Educational Resources at Post-Secondary Level in the Field of Environmental Modelling: An Exploratory Study. *Educ. Sci.* **2019**, *9*, 103. [CrossRef]
12. Cerezo-Narváez, A.; de los Ríos Carmenado, I.; Pastor-Fernández, A.; Yagüe Blanco, J.L.; Otero-Mateo, M. Project Management Competences by Teaching and Research Staff for the Sustained Success of Engineering Education. *Educ. Sci.* **2019**, *9*, 44. [CrossRef]
13. Malheiro, B.; Guedes, P.; Silva, M.F.; Ferreira, P. Fostering Professional Competencies in Engineering Undergraduates with EPS@ISEP. *Educ. Sci.* **2019**, *9*, 119. [CrossRef]
14. Neto, V. Eco design and Eco-Efficiency Competencies Development in Engineering and Design Students. *Educ. Sci.* **2019**, *9*, 126. [CrossRef]

15. Kaewunruen, S. Enhancing Railway Engineering Student Engagement Using Interactive Technology Embedded with Infotainment. *Educ. Sci.* **2019**, *9*, 136. [CrossRef]
16. Sena-Esteves, T.; Morais, C.; Guedes, A.; Pereira, I.B.; Ribeiro, M.M.; Soares, F.; Leão, C.P. Student's Perceptions Regarding Assessment Changes in a Fluid Mechanics Course. *Educ. Sci.* **2019**, *9*, 152. [CrossRef]
17. Maxworth, A. An Extended Constructive Alignment Model in Teaching Electromagnetism to Engineering Undergraduates. *Educ. Sci.* **2019**, *9*, 199. [CrossRef]
18. Hussain, S.; Gamage, K.A.; Sagor, M.H.; Tariq, F.; Ma, L.; Imran, M.A. A Systematic Review of Project Allocation Methods in Undergraduate Transnational Engineering Education paper. *Educ. Sci.* **2019**, *9*, 258. [CrossRef]
19. Viegas, C.; Marques, A.; Alves, G.R. Engineering Education addressing Professional Challenges. In Proceedings of the Seventh International Conference on Technological Ecosystems for Enhancing Multiculturality (TEEM'19), León, Spain, 16–18 October 2019; Miguel, Á.C.G., Francisco, J.R.S., Camino, F.L., Francisco, J.G.-P., Eds.; ACM: New York, NY, USA, 2019; pp. 51–53. [CrossRef]
20. MDPI–Education Sciences: Special Issue: Engineering Education Addressing Professional Challenge. Available online: https://www.mdpi.com/journal/education/special_issues/Engineering_Education_Addressing_Professional_Challenges (accessed on 15 November 2019).

© 2019 by the authors. Licensee MDPI, Basel, Switzerland. This article is an open access article distributed under the terms and conditions of the Creative Commons Attribution (CC BY) license (http://creativecommons.org/licenses/by/4.0/).

education sciences

MDPI

Article

Fostering Professional Competencies in Engineering Undergraduates with EPS@ISEP

Benedita Malheiro [1,*], **Pedro Guedes** [2,*], **Manuel F. Silva** [1,*] and **Paulo Ferreira** [3,*]

1   Department of Electrical Engineering, Polytechnic Institute of Porto, 4249-015 Porto, Portugal
2   Mathematics Department, Polytechnic Institute of Porto, 4249-015 Porto, Portugal
3   Department of Informatics, Polytechnic Institute of Porto, 4249-015 Porto, Portugalll
*   Correspondence: mbm@isep.ipp.pt (B.M.); pbg@isep.ipp.pt (P.G.); mss@isep.ipp.pt (M.F.S.); pdf@isep.ipp.pt (P.F.)

Received: 30 April 2019; Accepted: 21 May 2019; Published: 29 May 2019

check for updates

Abstract: Engineering education addresses the development of professional competencies in undergraduates. In this context, the core set of professional competencies includes critical thinking and problem solving, effective communication, collaboration and team building, and creativity and innovation—also known as the four Cs—as well as socio-professional ethics and sustainable development—referred in this paper as the two Ss. While the four Cs were identified by the associates of the American Management Association based on the needs of the society, professional associations, and businesses; this paper proposes the two S extension to ensure that future engineers contribute to the well-being of individuals and the preservation of life on Earth. It proposes a tangible framework—the 4C2S—and an application method to analyse the contributions made by engineering capstone programmes to the development of these core competencies in future engineering professionals. The method is applied to an engineering capstone programme—the European Project Semester (EPS) offered by the Instituto Superior de Engenharia do Porto (ISEP)—and a specific project case—EPS@ISEP Pet Tracker project developed in 2013, constituting, in addition, a road map for the application of the 4C2S framework to engineering capstone programmes. The results show that EPS@ISEP complies with the 4C2S framework.

Keywords: engineering education; capstone project; professional competencies; European Project Semester; 4C2S analysis framework

1. Introduction

Engineering education aims to prepare professionals to address the challenges of the future. This is a highly demanding goal since it not only implies a forecast of future needs but also anticipates scientific and technological advancement trends. Society (while beneficiaries), academia (while educators), and businesses (while employers) must work together to define the set of core competencies of future engineers.

The American Management Association (AMA) conducted in 2012 a critical skills survey among its corporate associates regarding the core professional competencies of the 21st century workforce. From the entrepreneurs' perspective, the 21st century business requires, beyond the basics of reading, writing, and arithmetics (the three R), skills such as *critical thinking and problem solving, effective communication, collaboration and team building*, and *creativity and innovation* (the four Cs) [1]. Specifically, the AMA defines *critical thinking and problem solving* (CTPS) as the ability to make decisions, to solve problems, and to take action as appropriate; *effective communication* (EC) as the ability to synthesise and transmit ideas both in written and oral formats; *collaboration and team building* (CTB) as the ability to work effectively with others, including those from diverse groups and with opposing points of view; and *creativity and innovation* (CI) as the ability to see what is not there and to make something happen.

However, according to Cohen and Grace [2], social responsibility should be seen as integral to the performance of engineers as individuals and of engineering as a profession. It involves thinking positively of social responsibility, avoiding harm, and consciously choosing to do good [2]. Since the engineering practice deals with the environment, professional ethics and behaviours, matters of health and safety, and discipline [3], additionally, this paper advocates the need to foster *sustainable development* (SD) and *socio-professional ethics* (SPE) (the two Ss) in the engineering practice. In this context, *sustainable development* corresponds to "development that meets the needs of the present without compromising the ability of future generations to meet their own needs" [4] and *socio-professional ethics* translate to the set of values governing the conduct of engineers in their role as professionals as well as individuals.

In the Bruntland report, the definition of sustainable development encompassed two key concepts: (*i*) the concept of needs, in particular the essential needs of the world's poor to which an overriding priority should be given, and (*ii*) the idea of limitations imposed by the state of technology and social organisation on the environment's ability to meet present and future needs [4]. Later, in 2004, the Barcelona Declaration reinforced the need to educate engineers for sustainable development. Specifically, it called for the multidisciplinary, systems-oriented, critical thinking, participative and holistic education of engineers [5]. More recently, in 2016, the United Nations 2030 Agenda for Sustainable Development, which defines 17 Sustainable Development Goals (SDG), officially came into force. While the SDG are not legally binding, governments are expected to take ownership and to establish a national framework for their achievement [6].

Ethics, as a whole, deals with the moral choices that are made by each person in his or her relationship with others, including those made while practising engineering. Engineering ethics encompasses the more general definition of ethics but applies it more specifically to situations involving engineers in their professional lives. Thus, engineering ethics is a body of philosophy indicating the ways that engineers should conduct themselves in their professional capacity [7]. Whether working on multinational project teams, navigating geographically dispersed supply chains, or engaging customers and clients abroad, engineering graduates encounter worlds of professional practice that are increasingly global in character [8]. In this context, the global engineer needs to be aware of effective ways to navigate these cultural differences, which is crucial for achieving their common goals [9]. This makes a strong case in favour of using multinational teamwork as a setup to learn socio-professional ethics. When students communicate, discuss, or organise their project work, they learn to respect the values of others and to develop global engineering competency as defined by Jesiek et al. [8].

The 4C2S framework proposes an enlarged set of competencies to analyse engineering education programmes. The framework allows the adoption of an evidence-based approach to identify how a programme contributes to the development of these competencies in future engineering professionals. The application of the approach is illustrated using an engineering capstone programme—the European Project Semester (EPS) offered by Instituto Superior de Engenharia do Porto (ISEP)—and a specific project case—the EPS@ISEP Pet Tracker project. The adopted method analyses the context and the team's learning journey using the framework, searching for evidences of the development of professional competencies within the process and timeline of the pet tracker project (activities, milestones, and deliverables).

The main contributions of this work are the proposed framework and method, which together attempt to identify and quantify evidences of the development of professional competencies within an engineering capstone programme.

In terms of organisation, this paper includes the following sections: Section 1 introduces the problem of the development of engineering professional competencies in undergraduates, the 4C2S analysis framework, and the structure of this document; Section 2 provides the perspectives of professional engineers associations from Europe, the United Sates of America (USA), and Australia regarding the most sought professional competences of future engineers; Section 3 presents similar capstone programmes; Section 4 details the approach followed to identify how a programme contributes to the development of these competencies in future engineering professionals; Section 5 describes, in general, the EPS and, in particular, the experiential active learning process of the EPS@ISEP from the perspective of the 4C2S framework; Section 6 analyses the case study of the development of the Pet Tracker project at the light of the proposed 4C2S framework; Section 7 discusses the outcomes; and Section 8 draws the conclusions.

2. Professional Competencies

Several national and international engineering accreditation organisations and engineering professional bodies around the world have disclosed criteria that should be followed by high education institutions graduating engineers:

European Network for Accreditation of Engineering Education (ENAEE) authorises the associated accreditation and quality assurance agencies in Europe to award to accredited engineering degree programmes the EURopean-ACcredited Engineer (EUR-ACE®) label, which is one of the European quality labels in higher education sponsored by the European Commission. The EUR-ACE® Standards and Guidelines for Accreditation of Engineering Programmes (EAFSG) are described in terms of student workload requirements, programme outcomes, and programme management [10]. ENAEE screens bachelor and master engineering programmes with reference to knowledge and understanding; (*ii*) engineering analysis; (*iii*) engineering design; (*iv*) research; (*v*) engineering practice; (*vi*) judgement making; (*vii*) communication and team-working; and (*viii*) lifelong learning. In particular, EASFG claims that exposing undergraduates to

- engineering practice enables the acquisition of "knowledge and understanding of the nontechnical—societal, health and safety, environmental, economic, and industrial—implications as well as critical awareness of economic, organisational, and managerial issues";
- judgement making enables the "ability to gather and interpret relevant data and handle complexity within their field of study; to inform judgements that include reflection on relevant social and ethical issues; to identify, formulate, and solve engineering problems in their field of study; as well as to manage complex technical or professional activities or projects in their field of study, taking responsibility for decision making";
- communication and team-working enables the ability "to communicate effectively information, ideas, problems, and solutions with the engineering community and society at large as well as to function effectively in a national and international context, as an individual and as a member of a team and to cooperate effectively with engineers and non-engineers";
- lifelong learning enables the ability "to recognise the need for and to engage in independent life-long learning and to follow developments in science and technology" at the bachelor level and "to engage in independent life-long learning and to undertake further study autonomously" at the master level.

Moreover, ENAEE details that master graduates need to demonstrate the ability to analyse, conceptualise, and solve unfamiliar problems, including the design of innovative analyses and problem-solving methods.

Engineering Council (UKEC) sets in the United Kingdom the overall requirements for the Accreditation of Higher Education Programmes in engineering in line with the UK Standard for Professional Engineering Competence. In order for an engineering degree to be accredited in UK, six broad areas of learning are analysed: (*i*) science and mathematics; (*ii*) engineering analysis; (*iii*) design; (*iv*) economic, legal, social, ethical, and environmental context; (*v*) engineering practice; and (*vi*) general skills [11]. According to the Engineering Council [11], Bachelor's degrees with honours are awarded to students who have demonstrated the following:

- *systematic understanding*, including the acquisition of coherent and detailed knowledge, and *conceptual understanding* to critically evaluate, make judgements, and frame appropriate questions to achieve a solution—or identify a range of solutions—to a problem;
- *awareness of the uncertainty, ambiguity, and limits of knowledge*;
- *ability to accurately apply methods and techniques of analysis and enquiry* to review, consolidate, and extend their knowledge and understanding and to initiate and carry out projects;
- *ability to communicate* information, ideas, problems, and solutions to both specialist and nonspecialist audiences.
- *ability to manage their own learning* and to make use of scholarly reviews and primary sources.

In terms of professional competencies, engineering graduates in the UK are expected to exhibit the following professional competencies: (*i*) *exercise of initiative and personal responsibility*; (*ii*) *decision-making in complex and unpredictable contexts*; and (*iii*) the *learning ability needed to undertake appropriate further training* of a professional or equivalent nature.

Accreditation Board of Engineering and Technology (ABET) in the United States defines a set of standards, called the Engineering Criteria 2000 (EC2000) [12], for engineering degrees. EC2000 shifted the basis for accreditation from inputs—what is taught—to outputs—what is learned—with the introduction of programme outcomes criteria [13]. The aim of these criteria is to ensure that students attain an *understanding of professional and ethical responsibility* as well as the broad education necessary to *understand the impact of technical solutions in a global, economic, environmental, and societal context*. Specifically, ABET specifies under Criterion 3 the so-called a–k list of student outcomes [14]. Moreover, according to a survey distributed to USA employers by the National Association of Colleges and Employers (NACE) [15], the three most important skills of an engineer today are (*i*) the ability to *communicate and work in teams*; (*ii*) the ability to *solve or troubleshoot problems in new or unfamiliar situations*; and (*iii*) *knowledge of a specific engineering* discipline.

Engineers Australia (EA) which performs in Australia the professional accreditation of engineering programmes, defines that engineering graduates must demonstrate at the point of entry to practice the following set of competencies [16]:

- *knowledge-oriented*—comprehensive and conceptual understanding; knowledge development and research; awareness of contextual factors impacting the engineering discipline; and understanding of the scope, principles, norms, accountabilities, and bounds of contemporary engineering practice;
- *application-oriented*—application of engineering methods, techniques, tools and resources and systematic engineering synthesis, design processes, and approaches to run and manage engineering projects;
- *profession-oriented*—ethical conduct and professional accountability, effective communication, creativeness, innovation and pro-activity, professional management and conduct, effective team membership, and team leadership.

The Engineers Australia summarises in Reference [17] its accreditation criteria, including the complete set of desired student educational outcomes.

Table 1 maps the desired engineering student skills identified by ABET, EA, UKEC, and ENAEE to the 4C2S framework. These employability skills correspond to the "ability to perform engineering related knowledge, skills, and personal attributes to gain employment, maintain employment, and succeed in the engineering field" [18].

Table 1. Worldwide perspectives of engineering professional skills and 4C2S competencies.

Competency	Body	Desired Professional Skill
Critical Thinking and Problem Solving	ABET	ability to understand the impact of engineering solutions (critical thinking) ability to identify, formulate, and solve engineering problems ability to recognise the need for and engage in life-long learning
	EA	ability to undertake problem solving, design, and project work ability to display critical reflection capacity for lifelong learning and professional development
	UKEC	ability to critically evaluate, make judgements, and frame appropriate questions to achieve a solution to a problem ability to manage their own learning
	ENAEE	ability to make judgements, identify, formulate, and solve engineering problems as well as to manage complex technical or professional activities ability to engage in independent life-long learning
Effective Comm.	ABET	ability to communicate effectively and work in teams
	EA	ability to display effective communication and pro-activity skills
	UKEC	ability to communicate with both specialist and nonspecialist audiences
	ENAEE	ability to communicate effectively with the engineering community and society
Collaboration and Team Building	ABET	ability to function on multidisciplinary teams
	EA	ability to assume effective team membership and team leadership
	UKEC	ability to work as a member of an engineering team and awareness of team roles
	ENAEE	ability to function in a national and international context, as an individual and as a member of a team, and to cooperate effectively with engineers and non-engineers
Creativity and Innovation	ABET	ability to apply knowledge creatively in order to solve a problem
	EA	ability to display effective creativeness, innovation, and pro-activity
	UKEC	ability to find creative solutions that are fit for purpose
	ENAEE	ability to design innovative analysis and problem solving methods (master level)
Sustainable Development	ABET	ability to consider economic, environmental, and sustainability constraints
	EA	ability to accommodate the economic and environmental responsibilities
	UKEC	ability to identify environmental and sustainability limitations
	ENAEE	ability to identify the environmental, economic, industrial, and managerial issues and to understand their implications
Socio-Professional Ethics	ABET	ability to work with professional and ethical responsibility
	EA	ability to accommodate social, cultural, ethical, legal, and political responsibilities as well as follow health and safety imperatives
	UKEC	ability to identify ethical, health, safety, security, risk, and intellectual property issues and to follow codes of practice and standards
	ENAEE	ability to inform judgements that include reflection on relevant social and ethical issues, taking responsibility for decision making

3. Engineering Capstone Programmes

The importance of grounding engineering education in real world experiences is highlighted by the National Academy of Engineering (NAE) in several publications [19,20]. In 2012, NAE reported on a selection of 29 capstone programmes offered by public and private universities and colleges in the USA, which successfully infused real world experiences into engineering undergraduate education [21]. The report features a diverse range of programmes (institution, category, scope, location, and longevity), potential implementation barriers, and strategies to overcome these barriers.

According to Hackman et al. [22], there are four main trends in engineering capstone projects: (*i*) the key role played by technology in the capstone experience (integration of technology into course administration, instructional methods, industry sponsorship and integration, and course evaluations); (*ii*) service-learning and community-based projects, in an attempt to provide real-world experiences while simultaneously providing a benefit to society; (*iii*) multidisciplinary projects, where student design teams are assembled from different majors or from different emphasis areas within a major; and (*iv*) the incorporation of sustainability principles into the capstone project, often in conjunction with community-based and service-learning projects. EPS@ISEP follows these trends (see Section 5).

The following engineering capstone programmes share similarities with the EPS@ISEP:

- Oladiran et al. [23] introduced the Global Engineering Teams programme, which adopts a multinational, intercultural, and geographically dispersed team-based approach. It tackled practical engineering problems, and each edition lasted for about six months between April and October. The groups in the programme were virtual teams consisting of students located in different countries and usually across multiple time zones, working in collaboration with industry partners. This programme, like EPS, implemented multinational, multidisciplinary teamwork and favoured real-world problems.
- Sheppard et al. [24] presented a two-semester pilot project at Stevens Institute of Technology to develop a systems engineering framework for multidisciplinary capstone design. It provided a series of workshops through the course of the capstone project to teach relevant systems engineering concepts in what approximates to a just-in-time mode. Interdisciplinary projects of significant scope were performed by teams of students from engineering and product architecture fields, working with external stakeholders and mentors. It was part of an initiative involving 14 institutions (including all the military academies), sponsored by the Department of Defense. The goal was to inculcate aspects of systems engineering into the education of students in all engineering disciplines through their major capstone project. The similarities with EPS included short intensive project supportive workshops, interdisciplinary projects, and multidisciplinary teams.
- Hackman et al. [22] described the new approach adopted by the School of Industrial and Systems Engineering at Georgia Tech to the capstone senior design course. The course structure creatively integrated internal and external resources for teaching, like EPS@ISEP, to promote business skills, soft skills, professionalism, and legal issues in an interdisciplinary, on-demand team-teaching format. Students formed teams and identified, scoped, and executed projects for real-world clients. The results showed that project quality and student nontechnical skills improved.
- Stanford et al. [25] reported on a capstone programme for civil engineering undergraduates employing a class-wide jigsaw approach and addressing community-based, sustainability-related problems. Results revealed that real-world projects with a focus on sustainability have a positive impact on students' critical thinking skills, leading to an increased knowledge of sustainability, and that open-ended problems with real project constraints could yield a uniquely beneficial learning experience without sacrificing the quality of student design or project deliverables. The pervasive concern with sustainability and the selection of real open-ended problems are common to EPS@ISEP.
- Palacin-Silva et al. [26] described a team-oriented capstone for software engineering undergraduates directed to the development of software services for sustainability. The course followed a collaborative learning approach, where students worked together to engineer a software project with the lecturer as a facilitator. The projects' challenge was to link Information and Communication Technologies to greening solutions, incorporating social, economic, and environmental concerns by involving computing, environmental sustainability, and citizen

engagement. This approach shares with EPS@ISEP the focus on sustainability-oriented design and sustainability problems.

4. Method

The 4C2S framework allows the adoption of an evidence-based approach to identify how a capstone programme contributes to the development of the identified competencies in future engineering professionals. To illustrate this empirical approach, the EPS@ISEP programme and the Pet Tracker project case are analysed from the 4C2S perspective.

The proposed method analyses the programme and the learning process of the students throughout the semester, searching for evidences of the development of critical professional competences, including the deliverables and the activities performed by the team. Specifically, the method looks for signs of professional behaviour and team work, e.g., the ability to meet deadlines, to define agendas, to lead meetings, to solve conflicts, to report and discuss findings, and to together reach a solution to an open-ended problem.

First, it identifies the skills related to the six core competencies of the framework (Table 1). Then, at the EPS@ISEP programme level, it matches these competencies against the aims (Table 4), learning process (Figure 1), and mandatory team deliverables (Table 3). Finally, at the project level, it analyses in detail the timeline of the Pet Tracker project to quantify the evidences of the development of the 4C2S competencies by the team. This timeline includes the scheduled activities—the project weekly meetings, the supportive module seminars, invited talks, presentation, and assessment events—and multiple milestones, involving the handing-in of different deliverables—Gantt chart, cardboard model, structural and control drawings, list of materials, components and providers, leaflet, brochure, report, poster, video, wiki, and prototype.

5. European Project Semester at ISEP

The European Project Semester was started in 1995 by Arvid Andersen [27]. It is a one semester student-centred international capstone programme currently offered by a group of European high education institutions called the EPS Providers as part of their student exchange programme portfolio more information available at http://www.europeanprojectsemester.eu/. While EPS Providers have the freedom to implement the programme with distinct flavours, they must comply with the EPS 10 Golden Rules listed at http://www.europeanprojectsemester.eu/. EPS provides an integrated framework for undertaking engineering capstone projects, which adopts a project-based learning methodology; focuses on teamwork; and exposes students to cultural, scientific, and technical diversity. In fact, it was designed for the capstone semester of engineering, product design, and business degrees. EPS aims to prepare future engineers to think and act globally [28].

The EPS is a package organised around one central module—the EPS project—and a set of complementary supportive modules. The project proposals refer to multidisciplinary real-world problems, i.e., draw on knowledge from diverse fields, and are open-ended, i.e., specify exclusively top-level requirements such as the compliance with the applicable directives/norms and the budget. According to the EPS 10 Golden Rules, the teams, which are composed of four to six students from different scientific backgrounds and nationalities, are fully responsible for the development of their projects [29]. Multicultural and multidisciplinary educational backgrounds contribute to product development and innovation, develop communication skills, and catalyse collaborative learning among team elements.

The EPS@ISEP programme—the EPS provided by the School of Engineering Instituto Superior de Engenharia do Porto of the Porto Polytechnic—welcomes engineering, business, and product design students since 2011. This 30 European Credit Transfer System Units (ECTU) programme is composed of six modules: Project, Project Management and Team Work, Marketing and Communication, Foreign Language, Energy and Sustainable Development, and Ethics and Deontology. Table 2 presents the programme syllabus. The 2 ECTU modules are project supportive seminars oriented towards the specificities of each team project.

Table 2. EPS@ISEP Syllabus.

Module	Acronym	ECTU
Project	PROJE	20
Project Management and Team Work	PRMTW	2
Marketing and Communication	MACOM	2
Foreign Language	PORTU	2
Energy and Sustainable Development	ESUSD	2
Ethics and Deontology	ETHDO	2

As far as project supervision is concerned, EPS@ISEP adopts a unique model where a panel of multidisciplinary expert advisers acts as a coaching and consulting committee. Concerning communication, the panel is aware that it interacts with students from diverse scientific and cultural backgrounds. Furthermore, in the weekly supervision meeting, only the topics previously specified by the team in the wiki agenda are discussed. Another very important aspect of the coaching methodology is the prompt feedback given to the students. Students meet with the panel once a week to discuss the topics the team previously posted in the wiki agenda.

During the semester, the teams maintain the project wiki and produce several deliverables, including the report, video, paper, manual, brochure, and proof of concept prototype. The structure and presentation of the deliverables are addressed in the communication seminar. The report structure (provided beforehand) includes as mandatory sections the introduction, state of the art, project management, marketing plan, sustainability, ethical concerns, conclusions, and bibliography. The marketing, ethical and deontological concerns as well as eco-efficiency and sustainability measures chapters are produced and refined within the corresponding complementary modules. Table 3 identifies the professional competencies developed with the production of these deliverables. In particular, the state of the art, project management, marketing plan, eco-efficiency measures for sustainability, and ethical and deontological concerns chapters not only report the team's corresponding studies but also specify the set of product requirements which were directly derived. The wiki is a key tool of the process. It is a collaborative work platform for team members and advisers, as well as the project show case, integrating the project plan, the weekly logbook, the report, and the deliverables areas.

Table 3. Mapping EPS@ISEP deliverables to 4C2S.

Deliverables	CTPS	EC	CTB	CI	SD	SPE
Interim Report		✓	✓		✓	✓
Interim Presentation		✓	✓	✓		✓
Final Report		✓	✓		✓	✓
Final Presentation		✓	✓	✓		
Paper	✓		✓	✓		
Poster		✓	✓	✓		
Leaflet		✓	✓			
User Manual	✓	✓	✓			✓
Video	✓	✓	✓	✓	✓	✓

Figure 1 illustrates this process.

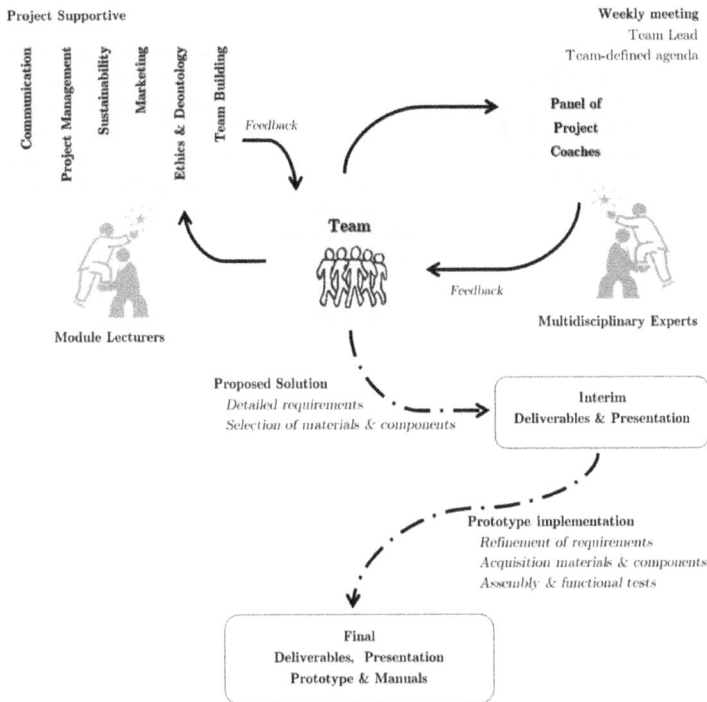

Figure 1. EPS@ISEP learning process.

The full process takes 15 weeks and includes the set of scheduled activities—the project weekly meetings, the supportive module seminars, invited talks, and global events—and multiple milestones. The global events, which involve lecturers, project coaches, students, and invitees, are the opening session in February (Week 1), the interim presentation in April (week 8), the final presentation in June (week 15), and the closing session, including certificate awarding and prototype hand-in. The milestones correspond to mandatory team and staff inputs. The different modules help drive the process:

Team Building | Week 1—Selection of the proposal and establishment of the Team Agreement, which defines the team's preferred conflict resolution method.

Project Management | Week 2–15—Definition of the activities and tasks, task allocation, and Gantt chart of the project (Week 2), followed by a continuous iterative refinement and adjustment cycle.

Communication | Week 2–15—Wiki maintenance, uploading of the interim deliverables (Week 7), presentation of interim outcomes (Week 8), refinement of interim deliverables (Week 10), uploading of final deliverables (Week 14), presentation of final outcomes (Week 15), and refinement (Week 16).

Sustainable Development | Week 1–7—Survey, application, and reporting of the relevant sustainable development practices and derivation of corresponding product requirements in the chapter "Eco-efficiency Measures for Sustainability".

Ethics and Deontology | Week 2–7—Study, selection, application, and reporting of applicable codes of ethics in order to derive product requirements in the chapter "Ethical and Deontological Concerns".

Marketing | Week 2–7—Research, definition, and reporting of the marketing plan of the proposed product and identification of resulting product requirements in the chapter "Marketing Plan".

Project Design | Week 2–7—Specification of the black box system diagrams and structural drafts (Week 3); analysis of the state of the art (reported in chapter "State of Art"); detailed specification of product requirements (Week 4), detailed system schematics, and structural drawings together with the card board scale model of the proposed solution (Week 5); and definition of the complete list of materials and components (Week 6).

Prototype Implementation and Operation (Week 9–15)—Procurement of components and materials, assembly, development, tests, and debugging.

Feedback and Assessment | Week 8 and Week 15—Interim Self and Peer (S&P) assessment (Week 7); discussion of the interim outcomes as a team (Week 8); feedback from peers (based on the S&P assessment) and staff, including improvement suggestions from staff; S&P assessment (Week 14); and discussion of the final outcomes as a team and individually (Week 15).

EPS@ISEP contributes to the development of (*i*) *critical thinking and problem solving* by specifying open-ended project proposals and by promoting inner team brainstorms and weekly meetings with a group of multidisciplinary coaches; (*ii*) *effective communication* by fostering intercultural and professional communication skills—the preparation of multiple textual and media deliverables during the semester and the definition of the agenda, leading, reporting the achievements, and writing the minutes of the weekly project meeting with the panel of advisers; (*iii*) *collaboration and team building* by working together in multicultural and multidisciplinary teams (defined according to the "EPS 10 Golden Rules" and the Belbin profiles); (*iv*) *creativity and innovation* by being expose to ill-defined open-ended problems, i.e., with very general requirements such as the budget, the applicable European Union (EU) directives, and the adoption of the International System of Units; (*v*) *socio-professional ethics* by tackling the related aspects of the project within the ETHDO module, with the writing of a dedicated report chapter, and the PROJE module, with the establishment of associated project requirements; and (*vi*) *sustainable development* by addressing all aspects related within the ESUSD and PROJE modules, resulting in the writing of a dedicated report chapter and the identification of related project requirements, respectively. Table 4 maps the EPS aims with the 4C2S framework.

Table 4. Mapping EPS@ISEP aims to 4C2S.

Aims	CTPS	EC	CTB	CI	SD	SPE
To train students in teamwork and to emphasise realistic and real-life situations	✓	✓	✓	✓	✓	✓
To demonstrate the ability to use modern design tools and techniques	✓	✓	✓			
To demonstrate the ability to plan and run a team-based project	✓	✓	✓	✓		
To show the ability to communicate clearly in writing (a proper project report) as well as by other means	✓	✓	✓	✓		

6. EPS@ISEP Pet Tracker Case Study

In the spring of 2013, a team of four students from Finland, Poland, Portugal, and Spain, with background education in Industrial Engineering and Management, Biotechnology, Electrical and Computer Engineering, and Computer Engineering chose to develop a pet tracker [30].

A pet tracker is a device used to monitor and track a pet, which can be used for different purposes, e.g., to follow hunting dogs or to verify the whereabouts of a domestic pet. While in terms of the determination of the location of the animal, these devices can use Global Navigation Satellite Systems (GNSS) or the Global System for Mobile Communications (GSM), in terms of the range of operation, i.e., communication with the owner, they rely either on the GSM/General Packet Radio Service (GPRS) network or on Radio Frequency (RF) transceiver pairs. The positioning accuracy corresponds typically, in the case of the usage of the GSM, to the cell size, i.e., from hundreds to thousands of meters while, in the case of the GNSS, to less than 5 m. While the main objective was to design and develop a pet tracking system, after analysing the state of the art, the team decided to add an activity monitoring feature for product differentiation. This feature allowed the owner to keep track of the pet's activity and to schedule the exercise it needed. According to the team, the product should "create a unique environment for the pet owner, where functionalities meet the client needs" [31]. Therefore, the final goal was to design, develop, and build a system allowing the pet owner not only to check the location of the pet but also to monitor and share on social media its daily level of activity.

The team, based on the market research, concluded that the Pet Tracker would be the first European pet activity tracking solution. This distinctive feature would make Pet Tracker unique in the European market. Regarding the particular market to address, the team decided first to aim for the Finnish market. The Finnish monthly median gross income was at the time 2776 € [32] with approximately 600,000 registered dogs, including approximately 450,000 pure-bred [33]. Finally, the students determined a final price for the Pet Tracker of 230 €, including fixed costs and a three month service. The monthly service fee packages contemplated by the team were (*i*) the 3-month package (19.99 €); (*ii*) the half-year package (34.99 €); and (*iii*) the full-year package (64.99 €).

The team considered the environmental, social, and economic components of sustainability. Concerning the environmental aspects, the students focused on the control and reduction of the materials and resources used, as well as on the waste produced. The goal was to select materials as durable and recyclable as possible, since using durable and recyclable materials reduces the ecological footprint and creates a better image for the company. They also planned to adhere to electronic device disposal services, promoting the reuse of components and the dispatch of leftovers to the appropriate disposal centres. Regarding the economic aspects of sustainability, among others, the team declared the need to have a continuous improvement/development process, including performance measurement, target setting, action taking, and result review. Such a continuous improvement process implies the development and adoption of quality, environmental, and risk management systems. The social sustainability perspective, i.e., the health and security of the customer and of the pet, was one of the most important concerns of the team. Consequently, the team became committed to designing a harmless and comfortable product both to humans and animals, i.e., the materials, components, and shape of the device must be safe for both the pet and the owner. Finally, the team analysed different codes of ethics and embraced the Fundamental Canons of the Code of Ethics for Engineers of the National Society of Professional Engineers [34]. The code was applied at different stages of the project. The team detailed the main concerns regarding the Pet Tracker from the marketing, environmental, safety and health, manufacturing, intellectual property, and liability viewpoints.

After taking into consideration these different project design dimensions, the team decided to create a product with the following requirements: (*i*) provide a Web interface for pet owners; (*ii*) display tracks using Google Maps; (*iii*) adopt open source technologies; and (*vi*) offer a light, small, wearable device for pets with on-board data storage, data download interface, and power autonomy of at least 48 h.

The proposed analysis framework was first applied to the team's wiki, which is available at http://www.eps2013-wiki2.dee.isep.ipp.pt/, specifically, to the project logbook. It found a total of 54 supporting evidences of the six core competencies. Figure 2 displays the results. As expected, the logbook provides multiple evidences of the different framework dimensions. Effective communication, critical thinking and problem solving, collaboration and team building, and socio-professional ethics are predominant. While the logbook provides few creativity and innovation and sustainable development evidences, multiple evidences of this competency can be found in the paper, report, and video of the team.

Figure 3 shows the EPS@ISEP timeline, providing an overall view of the global events, the mandatory teacher feedback on the left size and the team inputs on the right side. Furthermore, the corresponding Pet Tracker team inputs (deliverables) were analysed using the 4C2S framework. The central strip identifies the professional competencies developed by the team based on the evidences found in the deliverables, using a matrix format. The columns correspond, from left to right, to critical thinking and problem solving, effective communication, collaboration and team building, creativeness and innovation, sustainable development, and socio-professional ethics, whereas the lines identify milestones or deadlines. Whenever there is a clear evidence that the team activity induced the development of the corresponding professional competency, the cell is highlighted with the matching colour and a tick.

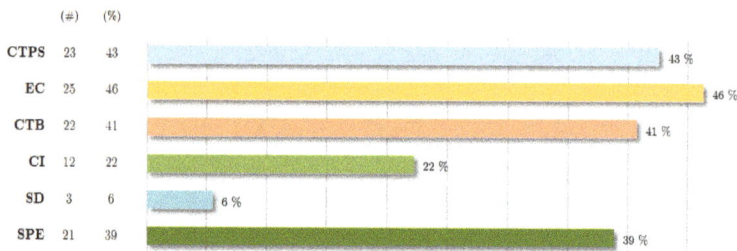

Figure 2. Evidences found in the weekly logbook.

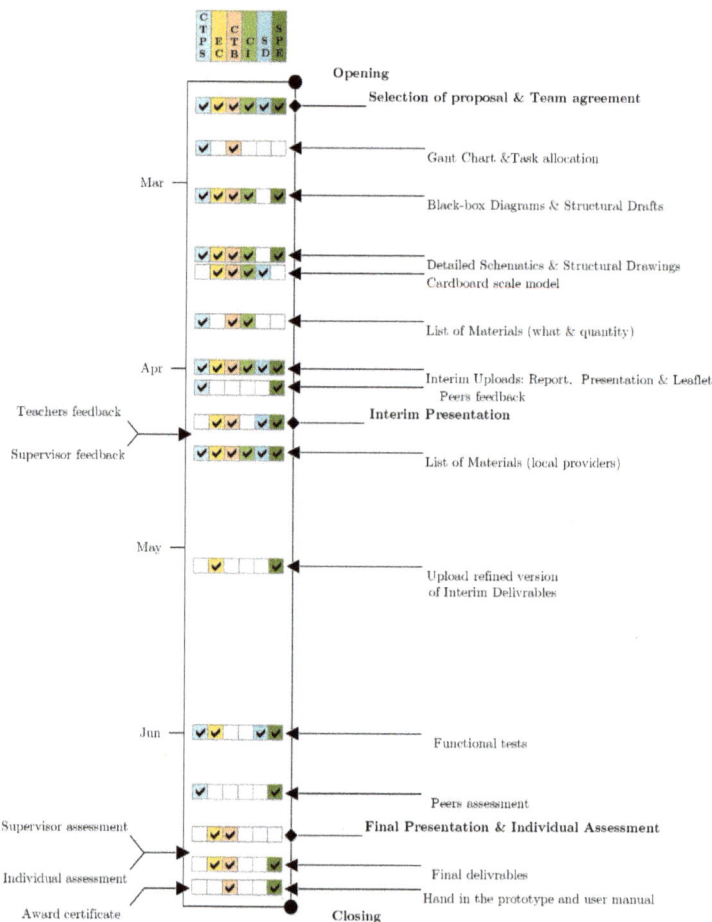

Figure 3. EPS@ISEP process timeline.

7. Discussion

Senior capstone programmes provide a comprehensive experience that helps students make the transition from academic life to professional life [22]. Not only do team-based projects encourage students to develop and improve employability skills, such as team work, problem solving, and intra-team communication [35] but also students undertaking real community-oriented projects report a significantly stronger development of work-ready skills [36]. EPS@ISEP not only develops these skills but also follows the trends identified by Hackman et al. [22]: relies on technology, selects service-learning and community-based multidisciplinary projects, and incorporates ethical and sustainability principles into the capstone project.

The 4C2S analysis framework is intended to identify evidences supporting the development by capstone programmes of the core professional competencies in future engineering graduates. Although it can be further detailed, it covers the critical competencies pinpointed by ABET, EA, UKEC, and ENAEE, namely, the scientific, personal, societal, ethical, and sustainability dimensions. These competency dimensions have been mapped into the set of desired engineering employability skills. According to the Royal Academy of Engineering [37], they comprise knowledge, wider employability skills, such as communication, team working, and organisational skills, and important professional behaviours such as ethics. Robinson [38] identified a growing demand for sustainability literacy as organisations across the public, private, and voluntary sectors sought graduates who could help them adapt their policies and practices to meet sustainability objectives. Moreover, the proposed extension of the 4Cs with 2Ss competencies is aligned with the employability agenda, which is about getting graduates to adapt to the new flexible workplace. On the one hand, Conlon [39] stated that "a narrow focus on the skills and values of individual students related to employability is not adequate to prepare them for the challenge of delivering sustainable and just engineering solutions" and that "students need to develop the capacity to situate their individual practice as engineers in its wider social context". On the other hand, for Robinson [40], "employability is about what makes for successful employment and about maximising good consequences for the individual and society; it is also intrinsically a good, and therefore not value-free".

Existing skills frameworks typically fall into the 4Cs competencies but fail to include the 2Ss. It is the case of the engineering employability skills framework of Yuzainee et al. [18] for Malaysian engineering graduates. It comprises three main groups—personal knowledge, personal attributes, and personal skills—and ten types of skills—communication, team work, lifelong learning, professionalism, problem-solving and decision-making, competency in application and practice, knowledge of science and engineering principles, knowledge of contemporary issues, engineering system approach, and competency in specific engineering disciplines.

The proposed 4C2S framework was applied to the learning process of a team of students throughout the capstone semester, searching for evidences of the development of the six critical professional competences. Specifically, it inspected the process timeline, the produced deliverables, and the attitude of the teams, identifying signs of professional behaviour and team work, e.g., the ability to meet deadlines, to define agendas, to lead meetings, to solve conflicts, to report and discuss findings, and to together reach a solution to an open-ended problem. It has been shown that the EPS@ISEP programme complies with the 4C2S framework, as depicted in Figures 2 and 3. On the one hand, the experiential learning techniques, such as those adopted by EPS@ISEP, provide various learning experiences such as critical thinking, collaborative learning, and peer evaluation [41]. On the other hand, the pervasive focus on sustainability and ethics imprints these concerns in the future engineers. The goal of the EPS programme is more ambitious than just expecting the students to implement prototypes (in this case the "Pet Tracker")—it also makes them contribute with their distinct visions of the problem to a common consensual solution. This process is not always easy, since at this educational level the students are not used to collaborating with peers from different nationalities (implying distinct cultural backgrounds) and from different study backgrounds (engineering students tend to think differently from business and product design students).

8. Conclusions

Engineering degrees have different focuses when it comes to the development of professional competencies in undergraduates. While the academic-oriented concentrate on the development of traditional hard competencies and the market-driven consider the 4Cs, this work proposes a holistic perspective, extending the 4Cs with socio-professional ethics and sustainability competencies (2Ss). This extension aims to reposition the engineering profession at the heart of the society by introducing well-being and well-doing as central goals of the engineering practice. In terms of contributions, this work proposes the 4C2S analysis framework, which derives from the needs identified by the society, professional associations, and businesses for Engineering education, as well as an application method for engineering capstone programmes. The adopted method searches for evidences of the development of the desired professional competencies in engineering undergraduates.

To illustrate the application of the framework, the EPS@ISEP engineering capstone implementation and the Pet Tracker project were described and analysed in the light of 4C2S. It involved the mapping of the aims, learning processes, and mandatory deliverables of the EPS@ISEP to the development of the six core competencies and the collection of related evidences from the deliverables and learning journey of the Pet Tracker team. The result shows that EPS@ISEP contributes to fostering the desired 4C2S competencies in engineering undergraduates, comprising the much sought combination of soft, hard, and character employability skills.

Author Contributions: Conceptualization, B.M.; Formal analysis, P.G.; Methodology, B.M.; Project administration, B.M.; Writing—original draft, B.M., Pedro Guedes, M.F.S. and P.F.

Funding: This work was partially financed by National Funds through the Portuguese funding agency, FCT—Fundação para a Ciência e a Tecnologia, within project UID/EEA/50014/2019.

Conflicts of Interest: The authors declare no conflict of interest.

Abbreviations

The following abbreviations are used in this manuscript:

4C2S	CTPS, EC, CTB, CI, SI, and SPE
AMA	American Management Association
ABET	Accreditation Board of Engineering Technology
CI	Creativity and innovation
CTB	Collaboration and team building
CTPS	Critical thinking and problem solving
EA	Engineers Australia
EC	Effective communication
ECTU	European Credit Transfer System Unit
ENAEE	European Network for Accreditation of Engineering Education
EPS	European Project Semester
ESUSD	Energy and Sustainable Development
ETHDO	Ethics and Deontology
GET	Global Engineering Teams
GPRS	Global Packet Radio Service
GNSS	Global Navigation Satellite System
GSM	Global System for Mobile Communications
ISEP	Instituto Superior de Engenharia do Porto
MACOM	Marketing and Communication
NAE	National Academy of Engineering
NSPE	National Society of Professional Engineers
PORTU	Foreign Language
PRMTW	Project Management and Team Work

PROJE   Project
RF   Radio Frequency
SD   Sustainable Development
SDG   Sustainable Development Goal
SPE   Socio-professional ethics
UKEC   United Kingdom Engineering Council
USA   United States of America

References

1. American Management Association. Executive Summary: AMA 2012 Critical Skills Survey. 2012. Available online: https://playbook.amanet.org/wp-content/uploads/2013/03/2012-Critical-Skills-Survey-pdf.pdf (accessed on 5 May 2019).
2. Cohen, S.; Grace, D. Engineers and social responsibility: An obligation to do good. *IEEE Technol. Soc. Mag.* **1994**, *13*, 12–19. [CrossRef]
3. Staniškis, J.K.; Katiliūtė, E. Complex evaluation of sustainability in engineering education: Case & analysis. *J. Clean. Prod.* **2016**, *120*, 13–20. [CrossRef]
4. Brundtland, G.; Khalid, M.; Agnelli, S.; Al-Athel, S.; Chidzero, B.; Fadika, L.; Hauff, V.; Lang, I.; Shijun, M.; de Botero, M.M.; et al. *World Commission on Environment and Development: Our Common Future (Brundtland Report)*; Technical Report; United Nations: New York, NY, USA, 1987.
5. Lozano, R.; Lukman, R.; Lozano, F.J.; Huisingh, D.; Lambrechts, W. Declarations for sustainability in higher education: becoming better leaders, through addressing the university system. *J. Clean. Prod.* **2013**, *48*, 10–19. [CrossRef]
6. United Nations. United Nations Sustainable Development Agenda. 2016. Available online: http://www.un.org/sustainabledevelopment/ (accessed on 5 May 2019).
7. Fleddermann, C.B. *Engineering Ethics*, 4th ed.; Prentice Hall: Upper Saddle River, NJ, USA, 2012; Volume 4.
8. Jesiek, B.K.; Zhu, Q.; Woo, S.E.; Thompson, J.; Mazzurco, A. Global Engineering Competency in Context: Situations and Behaviors. *Online J. Glob. Eng. Educ.* **2014**, *8*, 1–14.
9. Zhu, Q.; Jesiek, B.K. Engineering Ethics in Global Context: Four Fundamental Approaches. In Proceedings of the 2017 ASEE Annual Conference & Exposition, Columbus, OH, USA, 25–28 June 2017; pp. 1–12.
10. European Network for Accreditation of Engineering Education. EUR-ACE Framework Standards and Guidelines (EAFSG). 2015. Available online: http://www.enaee.eu/wp-assets-enaee/uploads/2012/02/EAFSG_full_nov_voruebergehend.pdf (accessed on 5 May 2019).
11. Engineering Council. The Accreditation of Higher Education Programmes: UK Standard for Professional Engineering Competence—Third Edition. 2014. Available online: http://www.engc.org.uk/engcdocuments/internet/Website/Accreditation%20of%20Higher%20Education%20Programmes%20third%20edition%20(1).pdf (accessed on 5 May 2019).
12. Accreditation Board for Engineering and Technology. Engineering Change: A Study of the Impact of EC2000. 2006. Available online: http://www.abet.org/wp-content/uploads/2015/04/EngineeringChange-executive-summary.pdf (accessed on 5 May 2019).
13. Accreditation Board for Engineering and Technology. Going Global Accreditation Takes Off Worldwide—2008 Annual Report For ABET Fiscal Year 2007–2008. 2008. Available online: http://www.abet.org/wp-content/uploads/2015/04/2008-ABET-Annual-Report-.pdf (accessed on 5 May 2019).
14. Accreditation Board for Engineering and Technology. Criteria for Accrediting Engineering Technology Programs, 2016–2017. 2017. Available online: http://www.abet.org/accreditation/accreditation-criteria/criteria-for-accrediting-engineering-technology-programs-2016-2017/ (accessed on 5 May 2019).
15. National Academy of Engineering. Engineering Technology Education in the United States. 2016. Available online: https://www.nap.edu/catalog/23402/engineering-technology-education-in-the-united-states (accessed on 5 May 2019).
16. Engineers Australia. Stage 1 Competency Standard For Professional Engineer. 2013. Available online: https://www.engineersaustralia.org.au/sites/default/files/content-files/2016-12/S02_Accreditation_Criteria_Summary.pdf (accessed on 5 May 2019).

17. Engineers Australia. S02 Accreditation Criteria Summary, Rev. 2. 2008. Available online: https://www.engineersaustralia.org.au/sites/default/files/shado/Education/Program%20Accreditation/110318%20Stage%201%20Professional%20Engineer.pdf (accessed on 5 May 2019).
18. Yuzainee, M.Y.; Zaharim, A.; Omar, M.Z. Employability skills for an entry-level engineer as seen by Malaysian employers. In Proceedings of the 2011 IEEE Global Engineering Education Conference (EDUCON), Amman, Jordan, 4–6 April 2011; pp. 80–85. [CrossRef]
19. National Academy of Engineering. *The Engineer of 2020: Visions of Engineering in the New Century;* The National Academies Press: Washington, DC, USA, 2004. [CrossRef]
20. National Academy of Engineering. *Educating the Engineer of 2020: Adapting Engineering Education to the New Century;* The National Academies Press: Washington, DC, USA, 2005. [CrossRef]
21. National Academy of Engineering. *Infusing Real World Experiences into Engineering Education;* The National Academies Press: Washington, DC, USA, 2012. [CrossRef]
22. Hackman, S.; Sokol, J.; Zhou, C. An Effective Approach to Integrated Learning in Capstone Design. *INFORMS Trans. Educ.* **2013**, *13*, 68–82. [CrossRef]
23. Oladiran, M.T.; Uziak, J.; Eisenberg, M.; Scheffer, C. Global engineering teams—A programme promoting teamwork in engineering design and manufacturing. *Eur. J. Eng. Educ.* **2011**, *36*, 173–186. [CrossRef]
24. Sheppard, K.G.; Nastasi, J.A.; Hole, E.; Russell, P.L. SE CAPSTONE: Implementing a Systems Engineering Framework For Multidisciplinary Capstone Design. In Proceedings of the 2011 ASEE Annual Conference & Exposition (ASEE 2011), Vancouver, BC, Canada, 26–29 June 2011; pp. 1–21.
25. Stanford, M.S.; Benson, L.C.; Alluri, P.; Martin, W.D.; Klotz, L.E.; Ogle, J.H.; Kaye, N.; Sarasua, W.; Schiff, S. Evaluating Student and Faculty Outcomes for a Real-World Capstone Project with Sustainability Considerations. *J. Prof. Issues Eng. Educ. Pract.* **2013**, *139*, 123–133. [CrossRef]
26. Palacin-Silva, M.V.; Seffah, A.; Porras, J. Infusing sustainability into software engineering education: Lessons learned from capstone projects. *J. Clean. Prod.* **2018**, *172*, 4338–4347. [CrossRef]
27. Andersen, A. The European Project Semester: A Useful Teaching Method in Engineering Education. In *Project Approaches to Learning in Engineering Education: The Practice of Teamwork;* de Campos, L.C., Dirani, E.A.T., Manrique, A.L., van Hattum-Janssen, N., Eds.; SensePublishers: Rotterdam, The Netherlands, 2012; pp. 15–28.
28. Andersen, A. Preparing engineering students to work in a global environment to co-operate, to communicate and to compete. *Eur. J. Eng. Educ.* **2004**, *29*, 549–558. [CrossRef]
29. Malheiro, B.; Silva, M.; Ribeiro, M.C.; Guedes, P.; Ferreira, P. The European Project Semester at ISEP: the challenge of educating global engineers. *Eur. J. Eng. Educ.* **2015**, *40*, 328–346. [CrossRef]
30. Borzecka, A.; Costa, A.; Fagerström, A.; Gasull, M.D.; Malheiro, B.; Ribeiro, C.; Silva, M.F.; Caetano, N.; Ferreira, P.; Guedes, P. Educating global engineers with EPS@ISEP: The "Pet Tracker" project experience. In Proceedings of the 2016 2nd International Conference of the Portuguese Society for Engineering Education (CISPEE 2016), Vila Real, Portugal, 19–21 October 2016; pp. 1–10. [CrossRef]
31. Borzecka, A.; Costa, A.; Fagerström, A.; Gasull, M.D. Pet Tracker—Final Report. European Project Semester, Instituto Superior de Engenharia do Porto. 2013. Available online: http://www.eps2013-wiki2.dee.isep.ipp.pt/doku.php?id=report (accessed on 5 May 2019).
32. Statistics Finland. Official Statistics of Finland (OSF): Structure of Earnings. 2011. Available online: http://tilastokeskus.fi/til/pra/2011/pra_2011_2013-04-05_tie_001_en.html (accessed on 5 May 2019).
33. The Finnish Kennel Club. Suomen Kennelliitto—Finska Kennelklubben—The Finnish Kennel Club. 2016. Available online: http://www.kennelliitto.fi/EN/kennelclub/kennelclub.htm (accessed on 5 May 2019).
34. National Society of Professional Engineers. NSPE Code of Ethics for Engineers. 2016. Available online: http://www.nspe.org/Ethics/CodeofEthics/index.html (accessed on 5 May 2019).
35. Keller, S.; Parker, C.M.; Chan, C. Employability Skills: Student Perceptions of an IS Final Year Capstone Subject. *Innov. Teach. Learn. Inf. Comput. Sci.* **2011**, *10*, 4–15. [CrossRef]
36. Gilbert, G.; Wingrove, D. Students' perceptions of employability following a capstone course. *High. Educ. Skills Work-Based Learn.* **2019**. [CrossRef]
37. Royal Academy of Engineering. Engineering Skills for the Future, The 2013 Perkins Review Revisited. 2019. Available online: http://www.raeng.org.uk/perkins2019 (accessed on 5 May 2019).
38. Robinson, Z. Linking employability and sustainability skills through a module on 'Greening Business'. *Planet* **2009**, *22*, 10–13. [CrossRef]

39. Conlon, E. The New Engineer: Between Employability and Social Responsibility. *Eur. J. Eng. Educ.* **2008**, *33*, 151–159. [CrossRef]
40. Robinson, S. *Ethics and Employability*; Learning & Employability, Series Two; The Higher Education Academy: Heslington, UK, 2005. Available online: https://www.qualityresearchinternational.com/esecttools/esectpubs/robinsonethics.pdf (accessed on 5 May 2019).
41. DeAgostino, T.H.; Jovanovic, V.M.; Thomas, M.B. Simulating Real World Work Experience in Engineering Capstone Courses. In Proceedings of the 2014 ASEE Annual Conference & Exposition (ASEE 2014), Indianapolis, IN, USA, 15–18 June 2014; pp. 1–13.

© 2019 by the authors. Licensee MDPI, Basel, Switzerland. This article is an open access article distributed under the terms and conditions of the Creative Commons Attribution (CC BY) license (http://creativecommons.org/licenses/by/4.0/).

education
sciences

MDPI

Article

Project Management Competences by Teaching and Research Staff for the Sustained Success of Engineering Education

Alberto Cerezo-Narváez [1,*], Ignacio de los Ríos Carmenado [2], Andrés Pastor-Fernández [1], José Luis Yagüe Blanco [2] and Manuel Otero-Mateo [1]

[1] School of Engineering, University of Cadiz, 11519 Puerto Real, Spain; andres.pastor@uca.es (A.P.-F.); manuel.otero@uca.es (M.O.-M.)

[2] School of Agricultural, Food and Biosystems Engineering, Technical University of Madrid, 28040 Madrid, Spain; ignacio.delosrios@upm.es (I.d.l.R.C.); joseluis.yague@upm.es (J.L.Y.B.)

* Correspondence: alberto.cerezo@uca.es; Tel.: +34-956-483-211

Received: 31 December 2018; Accepted: 18 February 2019; Published: 22 February 2019

check for
updates

Abstract: Projects have become an essential instrument for the success of universities. In a context of globalization and increasing complexity, they must sharpen their resourcefulness to face these challenges and adapt to this changing environment. To reach these objectives, they undertake a series of activities of a unique, concrete and temporary nature, not always technical but managerial ones. If universities work with people on projects in the production, transmission and dissemination of knowledge, then they link with society to solve its problems. For this reason, teaching and research staff (TRS) should promote a range of professional project management (PM) competences in different areas for the proper management of the projects in which they take part. Through a Delphi technique, a panel of twenty-four accredited teaching experts who are carrying out significant research and holding directive roles, measured the importance of acquiring and/or improving professional PM competences by their TRS. Consensus and stability reached after two rounds of consultation confirmed there are a series of crucial competences for the practice of relevant teaching and pioneer research. Results obtained are the basis for a gap plan that allows the TRS to participate in and/or lead university projects with greater self-confidence and personal motivation.

Keywords: project management; competences; engineering education; teaching and research staff

---

1. Introduction

In the university context, the TRS undertakes projects of diverse nature. However, despite their variety, their approaches are comparable to professional projects [1]. Firstly, there are research projects, with an administrative and documentary complexity normally proportional to their scope. Then, there are projects that arise from companies' needs, developed as collaborative initiatives through agreements, contracts or even industrial doctoral theses. Next, there are educational innovation and educative improvement projects. After that, there are entrepreneurship projects that help students engage themselves in real experiences. Finally, there is the university management itself, which covers many different types of projects, such as the design of internal management systems; the creation of research and educational innovation groups; the organization of faculties, technical schools, and departments; or even the assessment of degree programs, among many others. All these actions require stakeholder involvement, adapting their needs to specific requirements, and to carrying them out within planned schedules and budgets and with limited resources, considering risks and opportunities.

Teachers and researchers themselves usually formulate, manage, execute and/or evaluate different modalities of university projects. To do this, the TRS is not isolated but they are members of organizational structures, in which they administrate public and private resources, and engage both internal and external stakeholders to create, share and transfer knowledge to society. Even though research and educational innovation groups are autonomous organizations with their own strategies, governance, interests, culture and values, they insert within universities, supporting the formulation of projects. In this context, these groups manage their projects. Therefore, their members have to acquire and/or improve the necessary competences to work in projects.

Nevertheless, educational and research processes select and promote the TRS, without any consideration of management aspects, so they have to develop a range of competences in transversal areas, if they want to succeed in the projects in which they participate, as directors or as team members. Besides, if the recognition of their competences culminates in a formal process, including the issuance of a certificate of the competence possessed by an authorized institution [2], then the adaptability of the TRS is acquired, which facilitates their transferability in different contexts [3].

2. Background

2.1. Knowledge and Innovation Society

Universities are the center of the knowledge society. In fact, the link between universities and society and the organization of this around the abstract and universalized understanding of the world that universities provide are distinguishing features of the knowledge society [4]. The role of universities in stimulating innovation has long been accepted [5], bringing new ideas to society and being an integral economic engine. Figure 1 relates the sequence for the innovation process. Multiple factors and motivations that yield actions instigate innovation. These acts usually involve making inputs and driving research, development, production and distribution. Because of this, their outputs emerge as new knowledge and inventions, submit into outcomes as endeavors or enterprises, and impact on the promotion of more innovation activities.

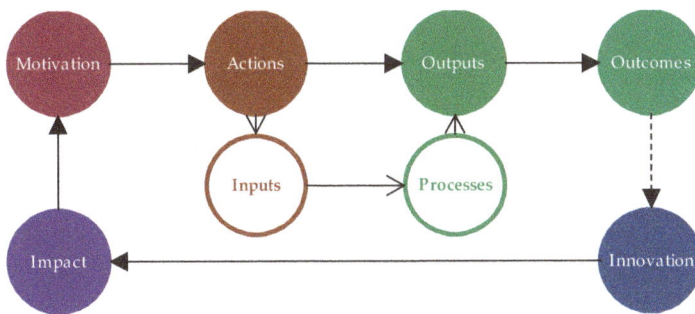

Figure 1. Innovation process in the knowledge society.

In this context, it is necessary to emphasize that knowledge adds value through its contribution to products, processes and people. Nevertheless, its management concerns with not only organizations but also universities, entities and public institutions. As shown in Figure 2, knowledge management includes its:

- Generation and development [6]
- Acquisition by an organization, identifying it from external environment and transforming it into an usable representation [7]
- Sharing, enhancing firms agility while improving stability [8]
- Capitalization, preserving and perpetuating the most critical one [9]
- Transfer, exchanging ideas, proofs and expertise and adding value [10]
- Application, maximizing organizational performance [11]

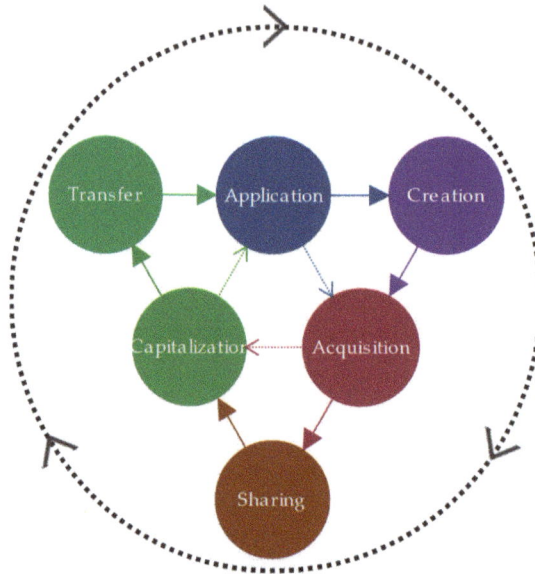

Figure 2. Knowledge management processes.

Furthermore, in a global and knowledge-driven economy, innovation is critical to competitiveness, long-term productivity growth, and prosperity. Thus, research bridges scientific discoveries and practical applications, as well as educates for giving skills to new generations, in order to convert knowledge into innovative products and services [12]. Consequently, several topics related to engineering appear in order to face challenges of the knowledge society. Under these circumstances, the address of complex problems that balance interdisciplinarity and commitment is the basis for the qualification of future engineers, once disciplinary skills are achieved [13].

2.2. Research Universities

In the knowledge and innovation society, research universities are key institutions for social and economic development. They are characterized by their global mission, research intensity, diversified funding, worldwide recruitment, increasing complexity, relationship with public administrations and industry, and global collaboration with other universities. These achievements are made through focusing on the discovery of new knowledge to develop the next generation of scholars, decision makers and entrepreneurs [14]. Moreover, research universities interact at different levels within the global market [15], including:

- Qualification possibilities which students benefit from
- Prestige associated with their publications
- Effectiveness and transferability of knowledge provided

On the other hand, the success of research universities depends on the TRS potential, funds in order to run, and a flexible structure. These properties allow them to succeed in different cultural and political contexts without sacrificing their autonomy and organizational vision. As summarized in Figure 3, research universities can be identified by [16]:

- Pioneer research, inspiring the TRS to generate new knowledge in a creative and useful way, as a stable driving force that ends connecting industry and university
- Relevant teaching, contributing to the strengthening and prestige of the institution in which it takes place, and being current, reflecting, suitable and not isolated

- Link with society, being significant to create substantial incomes in order to operate, and meeting social needs, at the forefront of progress, research and innovation

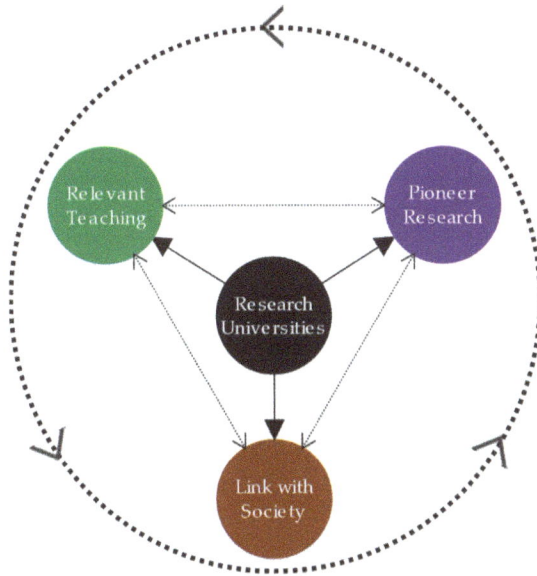

Figure 3. Intrinsic characteristics of research universities.

In summary, research universities serve the progress of society, solving its problems, threats, opportunities and/or needs [17]. Research universities are challenged to become the engine of transformation of society [18], recovering the original concept of the university as an institution of generation, tutelage and dissemination of knowledge [19]. International rankings such as the Academic Ranking of World Universities, Scimago Institutions Rankings, Center for World Global Universities Ranking, University Ranking by Academic Performance or the Ranking Web of Universities, among others, consider these aforementioned characteristics. In fact, if these are taken into account, then universities can lead to an improved position in rankings [20].

2.3. Project Management Competences

Projects have become omnipresent not only in economy but also in society [21]. However, they require an adjustment of organizational solutions, individual competences and changes in understanding their effects [22]. According to the Standish Group [23], almost twenty percent of all implemented projects are never finished, while forty-five percent are finished but with aberrations from their original goals, and only thirty-five percent can be described as efficiently implemented. In a context in which organizations face more and more challenges, it is necessary to find out what is needed to advance sustained and long-term solutions through increasingly more complex, fluid, and multicultural projects [24].

Additionally, the concept of competence in PM has been researched for many purposes, providing a detailed examination of its evolution [25] or explaining the role of knowledge in defining position descriptions [26]. However, the understanding and application of knowledge, tools and techniques recognized as good practices are not enough for effectively managing projects [27]. It also requires specific skills and general abilities.

Nevertheless, almost all PM standards are process-oriented. On the contrary, very few of them are competence-based, defining the specifications needed for a good performance of people in project environments [28]. While the first group of standards typically prescribes procedures and methods, ensuring organizations to have a universal approach in managing projects, the second one presents a wide spectrum of knowledge and skills that organizations need for success, holding people to perform tasks in projects [29]. From this perspective, the development of competences by personnel and maturity by organizations leads to the success of projects and related business [30].

On the other hand, the most extended and oldest PM associations worldwide are the International Project Management Association (IPMA) and the Project Management Institute (PMI). IPMA is a federation founded in 1965 and composed of seventy national associations. PMI is a professional membership association founded in 1969, with over half a million members and certification holders in one hundred eighty-five countries. Both IPMA [31–33] and PMI [34–36], as well as the International Standards Organization (ISO) 21500 [37] and European Union (EU) PM2 [38] standards, among others, guide their foundational standards, bodies of knowledge, methodologies, practical guides, baselines and frameworks focused on three approaches, as represented in Figure 4:

- Projects: Knowledge and practices to manage individual projects
- Organizations: Knowledge and practices to manage projects, programs and portfolios
- People: Development, counselling, registration and certification

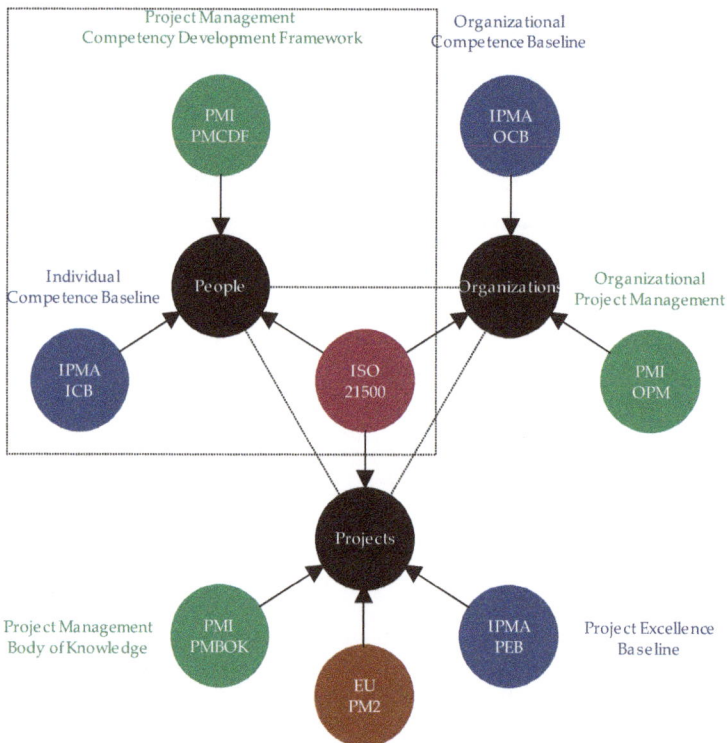

Figure 4. EU, IPMA, ISO and PMI project management approaches.

In the context of PM, competences include a set of abilities to mobilize knowledge, skills and resources to reach the expected performance in work, adding them economic and social value [39]. This combination of elements, related to work contexts (such as abilities, capabilities, expertise, experience, knowledge and skills), complement and integrate in conjunction with personal attributes (such as attitude, behavior, motivation, personality and personal values) [40].

The importance of both "hard" competences (relating to processes) and "soft" ones (dealing with people and their environment) is widely recognized in PM [41,42], but managers are ultimately responsible for balancing and optimizing their application. From this perspective, to learn about individual and organizational competences (and not only about technical ones) is critical, in order to complete the role transformation from engineers and other technicians to managers [43,44].

EU and ISO organizations and IPMA and PMI associations also group individual competences into three blocks:

- For IPMA ICB4 [31] and EU PM² [38]:   Perspective, practice and people
- For PMI PMBOK6 [36]:   Strategic-business, technical and leadership
- For PMI PMCDF3 [34]:   Knowledge, performance and personal
- For ISO 21500 [37]:   Contextual, technical and behavioral

Many researchers classify PM competences analogously as professional associations and international organizations:

- For Cheng, Dainty and Moore [45]:   Occupational, understanding and attitudinal
- For Crawford [46]:   Input, personal and output
- For Le Deist and Winterton [47]:   Social, functional and cognitive
- For Binkley et al. [48]:   Living in the world, tools for work and ways of thinking
- For Onisk [49]:   Compliance, professional and behavioral
- For Omidvar et al. [50]:   Contextual, job and personal
- For Teijeiro, Rungo and Freire [51]:   Instrumental, interpersonal and systemic
- For Chipulu et al. [29]:   Knowledge and expertise, managerial and personal traits

In summary, PM by competences, thanks to its intrinsic transverse and humanistic condition, covers management requirements in any sector [52], including research and education, which is considered as a professional sector as any other in this research. Within this framework, development of competences by the TRS allows achieving better project performance, thanks to growing motivation, better self-organization and reduced need for centralized control [53]. However, there are many ways of acquiring and improving them, depending on the organizational structure and its integrated management system, as well as on singularities and functions found in the organization chart in which they are located, as professional workers of the research and education sector.

2.3.1. Competences for PM practitioners

To manage projects requires a series of competences including interpersonal skills, technical abilities, cognitive aptitudes, abilities to understand both context and people and integrate leadership behaviors [54]. In fact, many studies highlight the impact of individual competences on project success [54–59]. In this regard, managerial skills and personal traits are critical to manage complex environments characterized by rapid changes and uncertainty [29].

In general, if organizations adjust their work arrangements to accommodate professional standards in PM, coordination is facilitated and performance is improved [60]. Among contrasted professional models, IPMA and PMI ones are internationally recognized in professional PM [61], and actually appear as more flexible and adaptive approaches than rigid frameworks [62].

On the one hand, the standard ICB 4 by IPMA [31] offers unique and role-specific competence development guidelines for improving project success, training and certifying practitioners. These professionals will probably work in disseminated environments with overlapping and conflicting stakeholder interests. In most cases, real-time data and performance management tools will shape them, too much information and not enough communication will challenge them, and their ability to deliver outcomes that align with short- and long-term strategies will judge them.

On the contrary, the standard PMCDF 3 by PMI [34] proposes the necessary specific skills and general management proficiencies required to domain for projects. At the same time, the standard PMBOK 6 by PMI [36] provides guidelines for managing individual projects and defines PM related concepts, such as methods, processes and practices.

To learn and train them, PMCDF and PMBOK by PMI standards focus on the end itself, from the premise that competences have a direct effect on performance. In opposition to them, the ICB by IPMA approach pays attention to the method itself, offering a series of proposals for individual development. Among them, self-study, peer-to-peer development, education and training, coaching and mentoring, and simulation and serious games are highlighted.

This approach based on international professional standards has demonstrated its utility for strategic projects (aligning objectives [63]), for educational projects (connecting teaching subjects with real-world problems [64]), and for research projects (driving the formation of the personnel [65]), influencing their effectiveness [66]. In Table 1, PM competences of the IPMA model are compared in pairs with PMI ones.

Table 1. Comparative between IPMA and PMI approaches.

IPMA ICB 4	Code	PMI PMCDF 3 and PMBOK 6
Perspective:		**Strategic and business management:**
Strategy	C01	Strategy and business
Governance, structures and processes	C02	Organizational process assets
Compliance, standards and regulations	C03	Organizational systems
Power and interest	C04	Politics and power
Culture and values	C05	Enterprise environmental factors
People:		**Personal:**
Self-reflection and self-management	B01	Managing
Personal integrity and reliability	B02	Professionalism
Personal communication	B03	Communicating
Relations and engagement	B04	Personality
Leadership	B05	Leading
Teamwork	B06	Being collaborative
Conflict and crisis	B07	Dealing with people
Resourcefulness	B08	Cognitive ability
Negotiation	B09	Getting things done
Result orientation	B10	Effectiveness
Practice:		**Technical:**
Design	T01	Tailoring
Requirements, objectives and benefits	T02	Goals and objectives
Scope	T03	Scope
Time	T04	Time
Organization and information	T05	Communication
Quality	T06	Quality
Finance	T07	Cost
Resources	T08	Human resources
Procurement and partnership	T09	Procurement
Plan and control	T10	Scheduling
Risk and opportunities	T11	Risks
Stakeholders	T12	Stakeholders
Change and transformation	T13	Integration
Select and balance	T14	Prioritization

2.3.2. Competences for Students

In education, cooperative project-based learning proposals and coworking competence-based training initiatives can introduce professional PM competences into theoretical educational frameworks [67,68]. In the case of the EU and Latin America, these approaches are broadly covered by [52]:

- Definition and Selection of Competencies (DeSeCo) Project by the Organization for Economic Co-operation and Development (OECD) [69] during the pre-university stage. It tries to instill that students assert rights and duties, communicate, conduct plans and projects, construct alliances, cooperate, empathize, make decisions, negotiate, recognize merits, resolve conflicts, are self-aware, suggest alternatives, support others, and take responsibilities.
- European Higher Education Area (EHEA) and the Latin America Academic Training (ALFA) Tuning Projects [70,71] during the university stage. They try to ensure that students analyze, appreciate diversity, are competitive, creative and critical, commit, communicate, lead, learn, make decisions, motivate, solve problems, synthesize, take initiative, and work as a team.

It is possible to organize both projects, as shown in Table 2, if knowing how to:

- Understand: Theoretical knowledge of academic fields
- Act: Practical application of knowledge to specific situations
- Be: Value as an integral element in social contexts

Table 2. Comparative among DeSeCo Project and Tuning Project competences.

OECD DeSeCo Project	ALFA amd EHEA Tuning Project
Use tools interactively:	**Instrumental:**
Reframe the problem	Problem solving
Learn from past actions	Applying knowledge in practice
Evaluate the value of information	Basic general knowledge
Analyze issues and interests	Working in international context
Understand of debate	Judgement of cultures and customs
Interact in heterogeneous groups:	**Interpersonal:**
Understand own interests	Criticism and self-criticism
Know rules and principles	Ethical commitment
Use communication skills effectively	Communication
Be empathetic	Appreciation of diversity
Make decisions	Leadership
Present ideas and listen to others	Teamwork
Manage emotions	Motivation
Suggest alternative solutions	Creativity
Negotiate	Cooperation
Identify action consequences	Initiative and entrepreneurial spirit
Act autonomously:	**Technical:**
Define projects and set goals	Project design and management
Prioritize needs and goals	Will to succeed
Have an idea of the system	Learning
Construct arguments	Research
Organize knowledge and information	Information management
Choose among available options	Concern for quality
Use technology	Elementary computing
Evaluate necessary resources	Working autonomously
Construct tactical alliances	Interaction with technical experts
Monitor progress	Organization and planning
Understand patterns	Analysis and synthesis
Identify areas of agreement	Working in heterogeneous teams
Access adequate information sources	Adaptation to new situations
Balance resources to meet goals	Decision making

First, the DeSeCo project helps young people develop as individuals and professionals in training projects that will last a lifetime, addressing complex demands by putting into action, in specific situations, psychological resources, skills and attitudes. After that, the Tuning projects seek to enable university students to prepare and carry out sufficiently and responsibly the tasks entrusted to them, as future professionals.

2.3.3. Competences in Engineering Education

In the field of engineering, companies require future engineers to have a wide range of competences that allow them to meet labor market expectations and to face successfully challenges that the changing world is promoting [72]. From this point of view, engineering education must add to main subject areas those competences that help them into entrepreneurial, environmental and social contexts and the understanding of professionals' characteristics [73–76]. Consequently, technical competences are no longer enough. Instead, the engineers' profile has to be based on the ability and willingness for learning, solid knowledge of basic natural sciences and good knowledge of any field of technology, in addition to general human and social values [77].

In the context of engineering education, three competence-based programs accreditation stand out for their dissemination and assurance of results:

- Accreditation Board of Engineering and Technology (ABET) accreditation   [78]
- Conceive → Design → Implement → Operate (CDIO) initiative   [79]
- European Accredited Engineer (EUR-ACE®) label   [80]

Comparative analyses among ABET, CDIO and EUR-ACE frameworks have been realized in the last years [72,77,81–86], concluding that there are many more similarities than differences, with all of them placing individual competences at the center of educational systems in which they have been implemented.

Firstly, the Accreditation Board of Engineering and Technology (ABET) was founded in 1980 in the United States by a series of member engineering societies. ABET shows the indispensable competences for engineers classifying them in two categories: hard skills (technical ones in nature) and professional ones (makers of real differences among professionals) [87], as exposed in Table 3. Besides, it underlines the importance of the competences needed for professional practice rather than emphasizing the curriculum [88]. Currently, almost eight hundred institutions of more than thirty countries are participating in ABET accreditation programs.

Table 3. ABET competences for engineering students.

Hard skills:
Apply knowledge of mathematics, science, and engineering
Design and conduct experiments, as well as to analyze and interpret data
Design a system, component, or process to meet desired needs within realistic constraints
Identify, formulate, and solve engineering problems
Use the techniques, skills, and modern engineering tools necessary for engineering practice
Professional skills:
Function on multidisciplinary teams
Understand of professionalism and ethical responsibility
Communicate effectively
Understand the impact of engineering solutions in a global, economic, environmental, and societal context
Engage in life-long learning
Know contemporary issues

Secondly, the Massachusetts Institute of Technology, in collaboration with Chalmers University of Technology, Linköping University and the Royal Institute of Technology, proposed in 2000 the CDIO initiative (Conceive → Design → Implement → Operate). The CDIO framework defines competences that students must own when completing their training as engineers, including not only generic, personal and interpersonal competences but also those that have traditionally been identified as typical of engineering [89], as shown in Table 4.

Table 4. CDIO competences for engineering students.

Technical knowledge and reasoning:	Interpersonal skills:
Basic science	Teamwork
Fundamental engineering	Communication
Advanced engineering	Foreign languages
Personal and professional attributes:	**CDIO in business and social contexts:**
Engineering reasoning	Social context
Problem solving	Business context
Experimentation and discovery	Conceive
Systemic thinking	Design
Personal attitudes	Implement
Professional skills	Operate

Likewise, the CDIO proposal masters a deep knowledge in fundamental techniques, leading the promotion of new products, processes and systems, and understanding the importance and strategic impact of research and technological development in society [90,91]. Currently, almost one hundred fifty institutions of almost forty countries are participating in CDIO accreditation programs.

Thirdly, the European Network for Accreditation of Engineering Education created the European Accredited Engineer (EUR-ACE®) system in 2006, after the Bologna Process. The EUR-ACE project formulates framework standards for the accreditation of higher education programs in engineering [92], as an entry route to the engineering profession. It has proved to be a powerful tool to improve both academic quality and relevance for the workplace [93]. Its main objective is to promote the quality of engineering graduates in order to facilitate their professional mobility and strengthen their personal and collective skills, as collected in Table 5. In summary, the EUR-ACE label assures that educational programs prepare graduates who are able to assume relevant roles in the job market [94]. Currently, more than three hundred institutions of more than thirty countries have accredited their engineering programs.

2.3.4. Competences for Workplace

Employability includes an array of technical and non-technical competences, encompassing knowledge, skills, expertise and even experience, to ensure that students are able to put them into practice, which is why educational stages must include them [95]. In the twenty-first century, organizations seek versatile individuals, even for entry-level jobs [96]. In this context, executives consider competences a very important attribute in labor applicants [97], becoming extremely important for job hires in many occupations [98], thanks to their potential role in maximizing business success [99].

Numerous studies have compared and compiled competences demanded by the labor market [100–103]. All of them agree on those related to or present in projects, such as communication, creativity, critical thinking, ethics, leadership, problem solving, professionalism, results orientation, self-management, or teamwork, among others. Shortly, candidates who add value with their competences have the ability to make a difference in obtaining and retaining the jobs for which they have been prepared [104].

Table 5. EUR-ACE competences for engineering students.

Knowledge and Understanding:
Scientific and mathematical principles underlying own engineering branch Key aspects and concepts of own engineering branch Forefront of own engineering branch
Engineering Analysis:
Identification, formulation and resolution of engineering problems using established methods Analysis of engineering products, processes and methods Selection an application of relevant analytic and modelling methods
Engineering Design:
Development and realization of designs to meet defined and specified requirements Use of design methodologies
Investigation:
Search of literature and use of data bases and other sources of information Design and conduction of appropriate experiments and interpretation of the data and drawing Workshop and laboratory skills
Engineering Practice:
Selection and use of appropriate equipment, tools and methods Combination of theory and practice to solve engineering problems Understanding of applicable techniques and methods, and of their limitations Awareness of the non-technical implications of engineering practice
Transferable skills:
Function effectively as individuals and as members of a team Use of diverse methods to communicate effectively with engineering community and society at large Awareness of health, safety and legal issues and responsibilities of engineering practice Commitment to professional ethics, responsibilities and norms of engineering practice Awareness of project management and business practices Engagement in independent and life-long learning

3. Objectives

The main objective of this research was to establish the importance of the TRS acquisition and improvement of professional PM competences in the university context. From the list of competences proposed by IPMA, the most important ones are the priority competences. If the development of professional PM practices had the necessary resources, then the universities would be closer to succeeding in the projects they have undertaken, and consequently they would contribute efficiently to society. The context of the research is schemed in Figure 5.

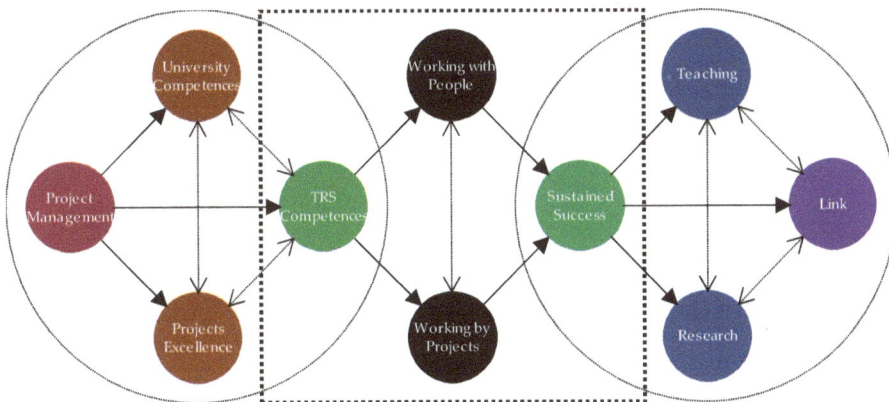

Figure 5. Framework of the research.

Knowledge society needs the transfer of new ideas to the market in order to make use of them. For this reason, universities become an essential economic driver and also play a crucial role in its construction, in terms of prosperity [105,106]. From an external perspective, universities confront these challenges undertaking projects that allow them to implement their strategies. At the same time, it requires a management system that responds to demands of adaptability, flexibility and availability, constituting itself as a device of change, adaptation and transformation [107,108].

PM tools and techniques can be applied to higher education sector [109,110], helping to affront challenges and barriers and improving its efficiency. If university projects are managed by PM methods to teach [111,112], research [113,114] and transfer [115,116], then the application of PM competences by the TRS can promote success in achieving objectives, providing value and generating synergies among institution members, universities, companies and social agents.

The IPMA ICB model focuses on people and helps to relate to a changing context and establish fundamental values to enhance society [64], incorporating human relations and social dynamics to the technical and technological dimensions. From a holistic point of view, the IPMA ICB approach is the most potentially applicable and useful professional PM framework at the university and incorporating sustained success principles [117,118]. However, other professional PM methodologies, such as PMI PMBOK or PMI PMCDF, which focus on processes, contribute the success of teaching and research projects, as flexible, open and transversal tools [119,120]. It can be noted that both are completely compatible. The former empowers the TRS and the latter establish a management system for their support.

4. Methodology

The purpose of this research was to emphasize the most relevant competences by the TRS for the development of projects undertaken in the university context. To achieve it, a Delphi technique was used. Figure 6 summarizes the process steps.

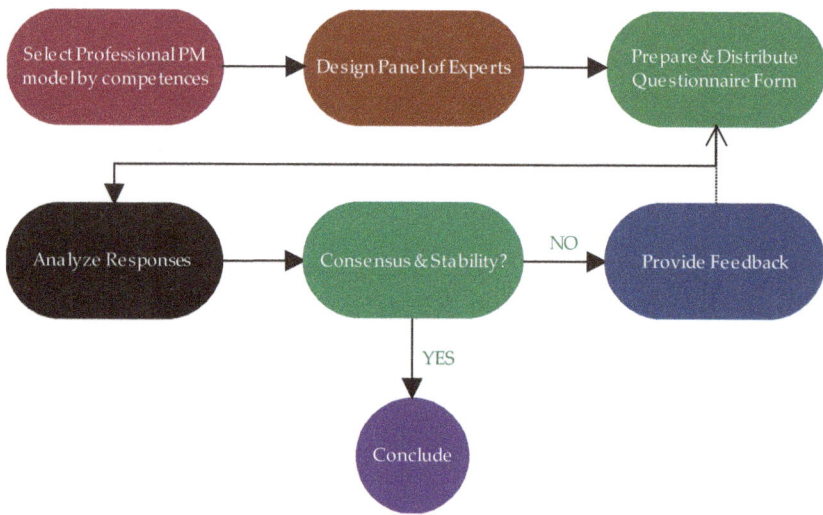

Figure 6. Research methodology.

The Delphi Technique is a prospective method for structuring an effective communication process that allows a group of individuals, as a whole, to deal with complex problems [121]. This process ends when the answers by a group of experts from a series of intensive questionnaires reach a reliable consensus and stability [122]. These iterations combine with controlled comments thanks to the provision of feedback from participants, who own expertise in the key area. At the same time, it is particularly useful to collect ideas on the specific topic and establish agreement to discover the underlying assumptions or perspectives among them, while avoiding the loss of its theoretical framework [123]. Once the process starts, the Delphi technique allows determining experts' points of arrangement, level of consensus and hierarchy of their importance.

The initial step to be done is the selection of experts [124]. That is, the Delphi technique has to be executed with the participation of individuals who have knowledge and competence in the study subject [123], as well as a deep understanding of the problem [125]. Therefore, the selection of the panel is one of the most critical actions of the process [126].

Thus, to be part of the initial sample, it was necessary that experts relate to engineering education, come from institutions where DeSeCo and Tuning projects are implemented and work in structures accredited (or in process of accreditation) by ABET, CDIO and/or EUR-ACE programs. Besides, with the aim of avoiding partiality, diversity and even lack of expertise, candidates had to comply additional requirements:

- Experience managing innovation educational and international research projects: At least five of each of them
- Experience in directive roles in universities: Faculty deans or directors of higher technical schools, departments, educational innovation and research groups Accredited relevant teaching experience: At least ten years of recognition (two quinquennia)
- Pioneer research at an international level: At least twelve years of impact and quality research (two sexennia)

Afterwards, it was necessary to design the research question that would be asked to the experts. The original research question was formulated in Spanish language and distributed among professors and researchers from Spanish and Latin American engineering schools, whereas both institutions and individuals had to meet the requirements set out above. The research question is translated exactly in the following terms:

"On a scale from 0 to 10, being 0 trivial and 10 essential, indicate the degree of importance that you grant to the acquisition and improvement of the following competences, by the teaching and research staff -TRS-, in the university context, in order to carry out the projects in which they participate, both for the practice of a relevant and sustainable teaching, especially in educational innovation projects, and for developing their research, in R+D+i projects, among others."

Then, the list of twenty-nine competences of the IPMA model and their definition (brief description, including purpose, knowledge needed and skills involved) was presented to experts asking them to rate their importance on a scale of 0–10, both for educational innovation projects on the one hand, and research projects on the other hand. Next, two stop criteria were predefined: achievement of consensus and stability. The fulfilling of the conditions imposed are [127,128]:

- Consensus was scored through the interquartile range (IQR):

  Definition: Difference between the third quartile (Q3) and the first one (Q1)
  Acceptance: Variation of equal or less than twenty percent

- Stability was calculated using the relative interquartile range (RIR):

  Definition: IQR divided into the second quartile (Q2)
  Admission: Variation within the twenty-five percent of the value range

Finally, it was necessary to evaluate answers obtained once reliable data were also validated. To this end and for every question, the results of the Delphi technique were distributed and categorized into five blocks through a double entry table, depending on their importance (much or little) and consensus (majoritarian or scarce) [129], as presented in Figure 7. According to it, crucial factors re those that have a high consensus and importance. Consequently, in this research, they must be the primary focus of attention for the acquisition, development and improvement of professional PM competences by the TRS.

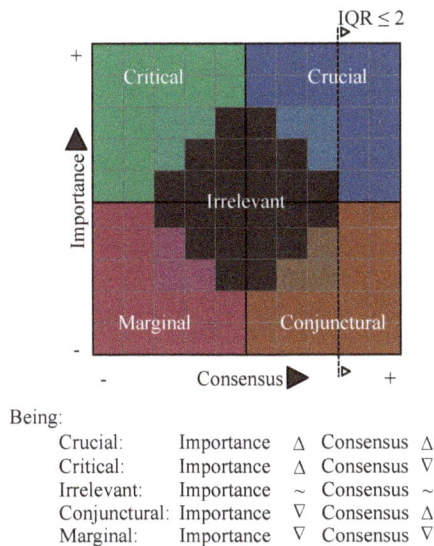

Being:
Crucial:	Importance	Δ	Consensus	Δ
Critical:	Importance	Δ	Consensus	∇
Irrelevant:	Importance	~	Consensus	~
Conjunctural:	Importance	∇	Consensus	Δ
Marginal:	Importance	∇	Consensus	∇

Figure 7. Competences' categorization according to importance and consensus.

5. Results

A total of twenty-four respondents meeting the requisites established participated from 25 January to 25 July 2018. The group of experts of the international academic community invited belong to sixteen universities from Spain and Latin America, as shown in Table 6.

Table 6. Location of experts.

Institutions	Country	Characteristics
Monterrey Institute of Technology	Mexico	
Polytechnic University of Catalonia (2)	Spain	
Polytechnic University of Madrid (4)	Spain	Polytechnic
Polytechnic University of Valencia (2)	Spain	
Salesian Polytechnic University	Ecuador	
Jaume I University	Spain	
Pontifical Catholic University	Chile	
University of Cadiz (2)	Spain	
University of Cordoba	Spain	
University of Granada (2)	Spain	Generalist
University of Huelva	Spain	
University of Oviedo	Spain	
University of Piura	Peru	
University of Seville (2)	Spain	
University of Valladolid	Spain	
National Distance Education University	Spain	Distance

Furthermore, expert applicants develop their activities in eighteen knowledge areas related to engineering, as summarized in Table 7.

Table 7. Knowledge areas of experts.

Knowledge Areas
Architectural projects
Business administration
Construction engineering (2)
Education science
Electrical engineering
Energy efficiency
Engineering projects (4)
Environmental technology
Industrial organization (2)
Inorganic chemistry
Market trading
Manufacturing processes (2)
Materials science
Mechanical engineering
Prospecting and mining
Rural development
Statistics and operational research
Structures

Finally, experience managing and leading teaching and research projects are briefed in Table 8, including the range of minimum and maximum values obtained.

Table 8. Experience and participation in teaching and research projects.

Teaching Experience	Teaching Projects	Research Experience	Research Projects
~ 24 years	~ 9	~ 20 years	~ 25
(from 11 to 39 years)	(from 7 to 33 ones)	(from 12 to 41 years)	(from 6 to 140 ones)

5.1. Competences for teaching

Consensus and stability were met after two rounds of consultation for teaching projects. In fact, only six of the twenty-nine elements of competence did not reach consensus in the first round (which are marked in **brown**). However, sixteen experts modified their opinions not only in those ones but also in fourteen other items in the second round (marked in **green** or **blue** if importance assigned increased and in **magenta** or **purple** if decreased), augmenting their respective consensus, as well as promoting a significant level of importance, as can be checked in Table 9.

Table 9. Results of Delphi panel for teaching.

Code	Round 1					Round 2				
	Importance			Consensus		Importance			Consensus	
	M [1]	SD [2]	Q2 [3]	IQR [4]	RIR [5]	M [1]	SD [2]	Q2 [3]	IQR [4]	RIR [5]
Contextual:										
C01	8.10	1.25	8	1	0.13	8.30	0.98	8	1	0.13
C02	7.10	2.05	7	2.75	0.39	7.60	1.35	7.5	1.75	0.23
C03	7.90	1.68	8	2	0.25	7.90	1.68	8	2	0.25
C04	7.25	1.89	7.5	1.75	0.23	7.40	1.43	7.5	1.75	0.23
C05	7.85	1.95	8	2	0.25	8.05	1.70	8	1.75	0.22
Behavioral:										
B01	7.85	2.30	8.5	2	0.24	8.00	2.05	8.5	2	0.24
B02	8.80	1.20	9	2	0.22	8.80	1.20	9	2	0.22
B03	9.05	0.89	9	1.75	0.19	9.05	0.89	9	1.75	0.19
B04	7.80	1.51	8	2	0.25	7.80	1.51	8	2	0.25
B05	7.75	1.59	8	1.75	0.22	7.75	1.59	8	1.75	0.22
B06	8.25	1.25	8.5	2	0.24	8.25	1.25	8.5	2	0.24
B07	7.90	1.45	8	2	0.25	7.90	1.45	8	2	0.25
B08	8.60	1.35	9	1	0.11	8.80	0.83	9	1	0.11
B09	7.35	1.39	7	1.75	0.25	7.35	1.39	7	1.75	0.25
B10	8.60	1.14	9	1.75	0.19	8.60	1.14	9	1.75	0.19
Technical:										
T01	8.40	0.99	8	1	0.13	8.55	0.89	8.5	1	0.12
T02	8.75	1.45	9	1.75	0.19	8.85	1.23	9	1.75	0.19
T03	7.35	1.53	7.5	1.75	0.23	7.50	1.47	8	1	0.13
T04	7.75	1.71	8	1.75	0.22	7.90	1.68	8	2	0.25
T05	7.35	1.31	7.5	1.75	0.23	7.50	1.19	8	1	0.13
T06	7.25	1.80	7.5	1	0.13	7.50	1.43	7.5	1	0.13
T07	7.00	1.78	6.5	2.75	0.42	7.45	1.19	7	1.75	0.25
T08	7.00	1.56	7	2	0.29	7.35	1.09	7	1.75	0.25
T09	5.70	2.30	6	2	0.33	6.20	0.77	6	1	0.17
T10	7.60	2.26	8	3.75	0.47	8.00	1.65	8	2	0.25
T11	6.75	1.62	7	2	0.29	6.65	1.31	7	1.75	0.25
T12	8.15	1.53	8	2.75	0.34	8.25	1.21	8	2	0.25
T13	7.35	1.84	7	2.75	0.39	7.60	0.99	7	1.75	0.25
T14	7.80	0.95	8	0.75	0.09	7.70	0.92	8	1	0.13
Average	**7.74**	**1.72**				**7.88**	**1.44**			

Note: [1] M: Mean; [2] SD: Standard deviation; [3] Q1: 1st Quartile (25th percentile); [3] Q2: 2nd Quartile (50th percentile), Q2 = Median; [3] Q3: 3rd Quartile (75th> percentile); [4] IQR: Interquartile Range, IQR= (Q3−Q1), ≤ 2.00 for consensus; [5] RIR: Relative Interquartile Range, RIR = (Q3−Q1)/Q2, ≤ 0.25 for stability.

5.2. Competences for Research

Consensus and stability were met after two rounds of consultation for research projects. In fact, only three elements of competence did not reach consensus in the first round (which are marked in brown) by the experts. However, eighteen experts modified their opinions in not only those but also in fifteen other items in the second round (marked in green or blue if importance increased), augmenting their respective consensus and promoting a significant level of importance, as can be checked in Table 10.

Table 10. Results of Delphi panel for research.

Code	Round 1					Round 2				
	Importance			Consensus		Importance			Consensus	
	M [1]	SD [2]	Q2 [3]	IQR [4]	RIR [5]	M [1]	SD [2]	Q2 [3]	IQR [4]	RIR [5]
Contextual:										
C01	9.10	0.85	10	1	0.11	9.15	0.81	9	1	0.11
C02	7.70	1.78	8	1.75	0.22	7.95	1.19	8	1.75	0.22
C03	8.60	1.14	9	1.75	0.19	8.60	1.14	9	1.75	0.19
C04	7.65	1.84	8	1.75	0.22	7.85	1.39	8	1.75	0.22
C05	8.10	1.74	8.5	2.75	0.32	8.25	1.29	8.5	2	0.24
Behavioral:										
B01	7.75	1.68	8	2	0.25	7.90	1.33	8	2	0.25
B02	8.95	1.00	9	2	0.22	8.95	1.00	9	2	0.22
B03	8.80	0.95	9	1.75	0.19	8.80	0.95	9	1.75	0.19
B04	8.20	1.28	8	2	0.25	8.20	1.28	8	2	0.25
B05	8.70	1.03	9	1.75	0.19	8.90	0.97	9	2	0.22
B06	9.00	0.92	9	1	0.11	9.05	0.94	9	1	0.11
B07	8.45	1.15	8	2.5	0.31	8.45	1.05	8	1	0.13
B08	8.90	0.79	9	1.75	0.19	8.90	0.79	9	1.75	0.19
B09	7.60	1.43	8	2	0.25	7.80	1.06	8	2	0.25
B10	8.85	1.04	9	2	0.22	8.85	1.04	9	2	0.22
Technical:										
T01	9.05	0.89	9	1.75	0.19	9.20	0.77	9	1	0.11
T02	8.90	1.33	9	1.75	0.19	9.05	1.10	9	1.75	0.19
T03	8.25	1.48	8	1	0.13	8.45	1.19	8	1	0.13
T04	8.45	1.36	8.5	1.75	0.21	8.45	1.36	8.5	1.75	0.21
T05	8.25	1.16	8	1	0.13	8.30	1.13	8	1	0.13
T06	7.90	1.68	8	1.75	0.22	8.15	1.14	8	1.75	0.22
T07	8.70	1.22	9	1.75	0.19	8.70	1.08	9	1	0.11
T08	7.95	1.67	8	1.75	0.22	8.20	1.06	8	1.75	0.22
T09	7.00	1.81	7	1.75	0.25	7.10	1.02	7	1.75	0.25
T10	8.15	1.95	8	2.75	0.34	8.50	1.36	8.5	2	0.24
T11	7.80	1.64	8	2	0.25	7.80	1.11	8	1.75	0.22
T12	8.15	1.60	8	1.75	0.22	8.45	1.05	8.5	1	0.12
T13	7.65	1.60	8	2.75	0.34	7.90	1.17	8	2	0.25
T14	8.20	0.77	8	1	0.13	8.20	0.89	8	1	0.13
Average	**8.30**	**1.45**				**8.42**	**1.18**			

Note: [1] M:Mean; [2] SD: Standard deviation; [3] Q1: 1st Quartile (25th percentile); [3] Q2: 2nd Quartile (50th percentile), Q2 = Median; [3] Q3: 3rd Quartile (75th percentile); [4] IQR: Interquartile Range, IQR= (Q3−Q1), ≤ 2.00 for consensus; [5] RIR: Relative Interquartile Range, RIR = (Q3−Q1)/Q2, ≤ 0.25 for stability.

6. Discussion of Results

The discussion of results consists of five subsections. First, the sample of experts is analyzed. Afterwards, the consensus and stability reached in answers is checked. Then, the importance given to each competence as isolated elements, both for educational innovation projects and for research ones, is assessed. Next, the network formed by the relationship among crucial competences, to highlight main nodes, is studied. Finally, the structure for a gap plan is developed.

6.1. Sample Representativeness

On the one side, EHEA or ALFA higher education areas insert universities to which experts belong, so traceability from the DeSeCo project to the Tuning one is ensured. In addition, these universities are either accredited or in the process of accreditation in a competence-based program (ABET, CDIO and/or EUR-ACE) in engineering. The entire sample complies with these institutional requirements. In relation to their physical location, twelve universities are Spanish and the other four are Latin American.

On the other side, the Delphi technique is a widely accepted method for gathering data, but only if respondents are within their domain of expertise [130]. Among knowledge areas related to main disciplines of engineering (construction, environment, industry, and technology), eighteen are included in the sample. Therefore, the different types of university projects related to engineering are widely represented. Finally, the size of the group is suitable if it is within the optimum range recommended, i.e. from six to thirty experts [131]. Twenty-four recognized experts composed the sample, thus it can be considered acceptable.

6.2. Validity and Reliability of Results

Validity and reliability increase transparency and decrease opportunities to insert researchers' bias in qualitative research [132]. Whereas reliability refers to the repeatability of findings, validity represents the truthfulness of findings [133]. Both refer to the consensus and stability of the results obtained [134].

IQR and RIR indexes measure consensus and stability, respectively. In this context, there is no need for experts to participate a third time, because variations were minimal after two rounds of consultation, thus results can be considered stable. At the same time, consensus was achieved. Therefore, for most questions, both IQR and RIR of the final round were lower than those of the initial one. In fact, convergence of responses was more common than divergence with more rounds [121].

However, the process reached consensus and stability for teaching projects in twenty-three competences in the first round, except C02, T07 and T10–T13. Analogously, the process achieved consensus and stability for research projects in twenty-six competences in the first round, except C05, T10 and T13. In summary, consensus and stability needed only one round of consultation on forty-nine of fifty-eight issues. Plan and control (T10) and Change and transformation (T13) were the competences with the least consensus and stability. On the contrary, Strategy (C01) and Select and balance (T14) were the competences with the most consensus and stability.

6.3. Grade of Importance

Once the process achieved the minimum level of consensus and stability thanks to the agreement of the experts, it was necessary to discuss the degree of importance obtained by each element of competence, for both educational innovation and research projects. If the importance was low, the element of competence was classified as conjunctural. On the contrary, if it was high, it was crucial.

In brief and to focus on the most crucial ones (those that realized a greater value of importance, once consensus and stability were ensured), a prioritized list of competences was extracted, as summarized in Table 11. However, it can be noted that all of them were crucial. Indeed, all elements of competence received more than half of the maximum score, for both types of projects.

The average score for the importance of professional PM competences based on the IPMA model for educational innovation projects was almost eight out of ten points. For educational innovation projects, two elements scored between 6–7 points, fifteen between 7–8 points, eleven between 8–9 points and one (B03) between 9–10 points. More in detail, technical competences reached an average of 7.5 points, contextual competences reached an average of 7.7 points and behavioral competences reached an average of 8.4 points.

Table 11. Prioritization of elements of competence according to their individual results.

Code	Element of Competence	Teaching	Research	Average	Priority
	Contextual:				
C01	Strategy	8.30	9.15	8.73	6
C02	Governance, structures and processes	7.60	7.95	7.78	22
C03	Compliance, standards and regulations	7.90	8.60	8.25	12
C04	Power and interest	7.40	7.85	7.63	26
C05	Culture and values	8.05	8.25	8.15	15
	Behavioral:				
B01	Self-reflection and self-management	8.00	7.90	7.95	19
B02	Personal integrity and reliability	8.80	8.95	8.88	3
B03	Personal communication	9.05	8.80	8.93	2
B04	Relations and engagement	7.80	8.20	8.00	17
B05	Leadership	7.75	8.90	8.33	10
B06	Teamwork	8.25	9.05	8.65	8
B07	Conflict and crisis	7.90	8.45	8.18	13
B08	Resourcefulness	8.80	8.90	8.85	5
B09	Negotiation	7.35	7.80	7.58	27
B10	Result orientation	8.60	8.85	8.73	7
	Technical:				
T01	Design	8.55	9.20	8.88	4
T02	Requirements, objectives and benefits	8.85	9.05	8.95	1
T03	Scope	7.50	8.45	7.98	18
T04	Time	7.90	8.45	8.18	14
T05	Organization and information	7.50	8.30	7.90	21
T06	Quality	7.50	8.15	7.83	23
T07	Finance	7.45	8.70	8.08	16
T08	Resources	7.35	8.20	7.78	25
T09	Procurement and partnership	6.20	7.10	6.65	29
T10	Plan and control	8.00	8.50	8.25	11
T11	Risk and opportunities	6.65	7.80	7.23	28
T12	Stakeholders	8.25	8.45	8.35	9
T13	Change and transformation	7.60	7.90	7.75	24
T14	Select and balance	7.70	8.20	7.95	20
Average		**7.88**	**8.42**	**8.15**	

Note: ▨ Eight most crucial competences.

By contrast, the average score in research projects was almost eight and a half points out of ten. For research projects, seven elements scored between 7–8 points, eighteen between 8–9 points and four (C01, B03, T01 and T02) between 9–10 points. Technical competences reached an average of 8.3 points, contextual competences reached an average of 8.4 points, and behavioral competences reached an average of 8.6 points.

Although the value obtained for educational innovation projects was almost 90%, in the case of research projects, the importance rose to almost 85%. Besides, there is a need to emphasize the relevance of behavioral competences for both educational innovation projects and research ones. However, as shown in Table 11 and considering all possible situations in a university context, between the most valued competence and the eighth, there was the same difference as between the eighth and the ninth, which implies that these competences make up the first gap.

6.4. Net of competences

In practical project situations, elements of competence are not isolated, because they are related each other. For that reason, the individual value of their importance should not be the unique criterion for their assessment. As competences are trained, performance is achieved not only by these elements but also by those with which they are related, contributing to each other's improvement [31]. Table 12 compiles the relationships among the eight most crucial competences, from the basis of the proposal of the IPMA ICB 4 model [31]. These relationships are multi-lateral, but being important enough for providers and receivers (establishing strong relations) or only for one of them (establishing weak or medium relations between providers and receivers).

Table 12. Relationships among crucial elements of competence.

Code		C01	B02	B03	B06	B08	B10	T01	T02
C01	Strategy	-							
B02	Integrity and reliability		-						
B03	Communication			-					
B06	Teamwork				-				
B08	Resourcefulness					-			
B10	Result orientation						-		
T01	Design							-	
T02	Requirements and objectives								-

Note: : Weak relationships; : Medium relationships; : Strong relationships.

Based on the relationships from Table 12 and according to the influences they exert on each other, as shown in Figure 8, the competence Result orientation (B10) is the center of the net of crucial elements of competence. It was the most relevant, having a relationship with the seven other ones. Next, Resourcefulness (B08) is highlighted, with six relations. This is followed by Communication (B03) and Requirements and objectives (T02), with five each. Then, Teamwork (B06), Integrity and reliability (B02) and Design (T01) with four and Strategy with three relations, the most isolated.

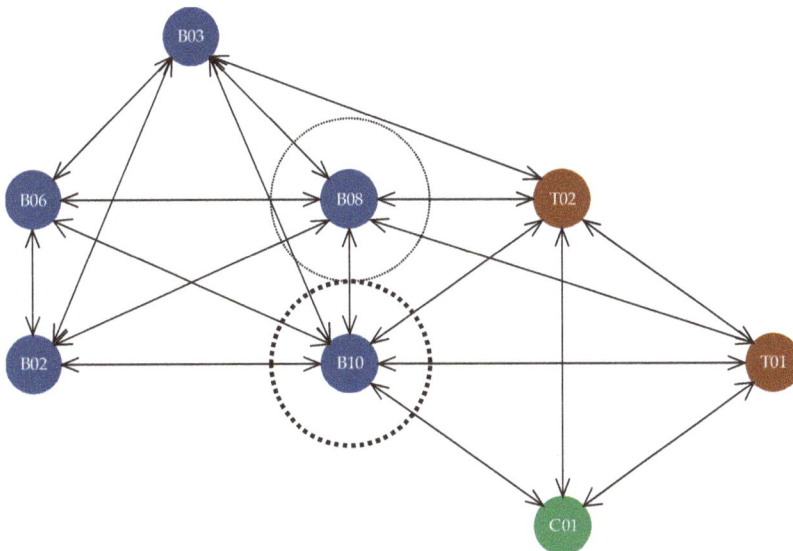

Figure 8. Net of essential elements of competence.

6.5. Basis for a Gap Plan

Once the most crucial elements of competence were identified and prioritized, isolated and together, it was necessary to lay the groundwork for their acquisition, development and improvement by the TRS. The following indicators can be used for their implementation, according to the guidelines of the IPMA ICB 4 model [31], with the help of the PMI PMCDF 3 and the PMI PMBOK 6 frameworks [34,36]:

- The competence C01 (Strategy) ensures the correlation between objectives and goals with the mission of the university. To do this, it is necessary to identify and exploit opportunities for influencing at the university strategy; develop and ensure the ongoing validity of its justification; and determine, assess and review critical success factors and key performance indicators.
- The competence T01 (Design) integrates demands, desires and influences, drafting how resources, funds, benefits, risks and opportunities, deliveries, priorities and urgencies are considered and deriving the proper approach to guarantee success. This requires acknowledging, prioritizing and reviewing success criteria; applying and exchanging lessons learned; determining complexity and its consequences for the approach; and selecting, if possible, the overall PM approach.
- The competence T02 (Requirements and objectives) deals with objectives, benefits, deliverables, requirements and outcomes and how they relate to each other. This implies defining and developing goals hierarchy; identifying and analyzing needs and expectations; and prioritizing and deciding on acceptance criteria.
- The competence B02 (integrity and reliability) builds integrity, reliability and responsibility from ethics, commitment and trust. For this, it is necessary to acknowledge and apply ethical values to both decisions and actions derived; to promote the viability and consolidation of outputs and outcomes; to take responsibility for own decisions and actions; to act, take decisions and communicate in a consistent way; and to complete tasks thoroughly in order to build confidence with others stakeholders.
- The competence B03 (Personal communication) exchanges adequate information and delivers it with precision and coherence to relevant parties. Because of it, there is clear and structured information to verify their understanding; to facilitate and promote open communication; to choose communication styles and channels to meet audience needs; to communicate effectively with virtual teams; and to employ humor and perspective when appropriate.
- The competence B06 (Teamwork) brings people together to realize common goals, building a productive team by forming (selecting right members), supporting (promoting orientation) and leading (managing the team). This involves selecting and building the team; promoting cooperation and networking between team members; supporting, facilitating and reviewing the development of members team; empowering teams by delegating tasks and responsibilities; and recognizing errors to facilitate learning from mistakes.
- The competence B08 (Resourcefulness) facilitates applying ways of thinking for the definition, analysis, prioritization, finding alternatives for, dealing with and solving challenges and problems, in order to manage better and more effective approaches. This means stimulating and supporting an open and creative environment; applying conceptual thinking to define strategies and analytic techniques for the analysis of situations, data and trends; and promoting creative techniques to find alternatives and solutions and a holistic view of the context to improve decision-making.
- The competence B10 (Result orientation) prioritizes resources to overcome problems, challenges and obstacles in order to focus on productivity, as a combination of effectiveness and efficiency. This implicates evaluating all decisions against their impact on success and objectives; balancing needs and means to optimize outcomes and success; creating and maintaining a healthy, safe and productive working environment; promoting projects, their processes and outcomes; and delivering results and getting their acceptance.

7. Conclusions

Projects are essential by their contribution to the sustained success of universities. In a scenario in which the production of knowledge results from scientific research, its transmission takes place through education and training, its dissemination thanks to information and communication technologies and its exploitation by innovation, universities postulate as the engine of social and economic change. From this point of view, research universities, working on projects with people in relevant teaching and pioneering research, link with society to influence a responsible development [15,135]. All these objectives, targets and goals, once formulated, lead to a series of projects.

Accordingly, to survive in this competitive environment, universities look for a competitive advantage, emphasizing the availability of potential competent staff, for which they can make a remarkable effort by increasing their competence. In engineering education, the TRS is a specialist at the highest level in engineering (science, technology and business, among others), which involves the capacity and investigative habits that allow them to approach and expand the frontiers of their branch of knowledge [136].

Frameworks based on competences in higher education have been successfully implemented [67,137–139]. At this point, this research joins previous ones demonstrating that professional PM competences help to improve in a sustained manner the results of university projects undertaken by the TRS. Although projects tackled in an unstructured way can succeed, the chances of repeating it significantly increase if the university structures create the appropriate conditions for the TRS. If teachers and researchers are university professionals, then they are equivalent to other sector practitioners, for whom project-based approaches are successfully operating. In this case, they can compare with each other to establish synergies.

Among projects that the TRS has to face, two of them stand out: educational innovation projects and research ones. To manage them, the TRS not only has to deal with the technical processes in which they are implied, but they also have to organize and coordinate, collaborate and cooperate as a team.

The twenty-four experts consulted agree that PM competences help the TRS to address their teaching and research, leading to a successful conclusion of their projects, based on a responsible formulation of objectives and management of the necessary activities. The Delphi panel showed that the acquisition and improvement of professional PM competences by the TRS is essential in order to engage projects in which they participate towards the achievement of results.

Among the twenty-nine elements of competence of the IPMA ICB 4 model (compatible with PMI PMBOK 6 and PMCDF 3 models), eight of them stood out, in consensus and stability (as valid and reliable sources), importance (as isolated elements) and influence (as interconnected nodes). Therefore, they are the necessary core to manage projects in the university community. These crucial competences are: Strategy from the contextual domain, Design and Requirements and objectives from the technical domain and Integrity and reliability, Personal communication, Teamwork, Resourcefulness and Result orientation from the behavioral domain.

In the university context, experts stress three elements of competence from the professional PM discipline for the proper resolution of projects. Strategy competence (C01) encompasses the formal justification of projects objectives and the establishment of long-term goals [140,141]. Design competence (T01) addresses the design, development, implementation and maintenance of an approach that takes into account all formal and informal factors that help to success of university projects [142,143]. Requirements and objectives competence (T02) establishes the relationship between what stakeholders (students, colleagues and institutions, among others) want to achieve and what projects are going to accomplish [144,145].

Reciprocally, the acquisition and improvement of PM competences by the TRS for carrying out the projects in which they are involved, both for the practice of a relevant teaching, especially in educational innovation projects, and for developing their research, predominantly in research, development and innovation projects, helps to ensure committed results. To manage them (leading people and administrating available resources), five elements of competence stood out, according to the experts.

Personal integrity and reliability competence (B02) enables making consistent decisions, taking congruous actions and behaving consistently in the projects undertaken [146,147], whereas Personal communication competence (B03) describes the essential aspects of an effective communication [148,149]. Teamwork competence (B06) promotes a team orientation, and effectively manages a team [150,151]. Resourcefulness competence (B08) effectively handles uncertainty and changes by searching for new, better and more effective solutions [152,153]. Results orientation competence (B10) enables focusing on the agreed outputs and outcomes and driving the success [154,155].

However, it is necessary to mention an observation. This research developed thanks to the collaboration of the experts panel, who come from Spain and four Latin American countries, which may be a limitation to the research findings. Nevertheless, the choice of experts, who are carrying out their work in centers in which the development of engineering competence-based accredited (or in the process of accreditation) programs (as ABET, CDIO or EUR-ACE) seeks to mitigate this potential cultural effect and can therefore be exported to other contexts where the Tuning project (from EHEA or ALFA areas) is implemented.

As a continuation of this research and future line of action, checking the degree of maturity in PM of the TRS that intervenes in educational innovation and research projects, using key competence indicators, is the following step. After that, with the measure and evaluation of the maturity level in PM of the TRS done, university structures can accordingly implement a customized breeding procedure, from the basis of the gap plan proposed, as the next step to develop the acquisition and improvement of their PM competences.

Author Contributions: Conceptualization, A.C.-N., I.d.l.R.C. and A.P.-F.; methodology, A.C.-N.; software, A.C.-N.; validation, A.C.-N., J.L.Y.B. and M.O.-M.; formal analysis, I.d.l.R.C. and A.P.-F.; investigation, A.C.-N.; resources, I.d.l.R.C. and A.P.-F.; data curation, A.C.-N.; writing—original draft preparation, A.C.-N.; writing—review and editing, A.C.-N., J.L.Y.B. and M.O.-M.; visualization, A.C.-N.; supervision, I.d.l.R.C. and A.P.-F.; project administration, A.C.-N., I.d.l.R.C. and A.P.-F.; funding acquisition, A.C.-N., A.P.-F. and M.O.-M.

Acknowledgments: To the GESPLAN Group of Technical University of Madrid, the INTELPREV Group of the University of Cadiz and the own Plan of aid for the promotion of research at the University of Cadiz.

Conflicts of Interest: The authors declare no conflict of interest.

References

1. Miterev, M.; Engwall, M.; Jerbrant, A. Exploring program management competences for various program types. *Int. J. Proj. Manag.* **2016**, *34*, 545–557. [CrossRef]
2. Zúñiga, F.V. *Competencias Clave y Aprendizaje Permanente*; Herramientas para la transformación N°26; CINTERFOR: Montevideo, Uruguay, 2004; ISBN 978-9290881698.
3. Guerrero, D.; De los Ríos, I. Learning model and competences certification in the project management scope: An empirical application in a sustainable development context. *Procedia Soc. Behav. Sci.* **2012**, *46*, 1297–1305. [CrossRef]
4. Frank, D.J.; Meyer, J.W. University expansion and the knowledge society. *Theory Soc.* **2007**, *36*, 287–311. [CrossRef]
5. Hawkins, R.W.; Langford, C.H.; Sidhu, K.S. University research in an "innovation society". In *Science, Technology and Innovation Indicators in a Changing World: Responding to Policy Needs*; OECD Publishing: Ottawa, ON, Canada, 2007; pp. 171–191, ISBN 978-9264039667.
6. Izunwanne, P.C. Developing an understanding of organisational knowledge creation: A review framework. *J. Inf. Knowl. Manag.* **2017**, *16*, 1750020. [CrossRef]
7. Holsapple, C.W.; Jones, K.; Leonard, L.N.K. Knowledge acquisition and its impact on competitiveness. *Knowl. Process Manag.* **2015**, *22*, 157–166. [CrossRef]
8. Intezari, A.; Taskin, N.; Pauleen, D.J. Looking beyond knowledge sharing: an integrative approach to knowledge management culture. *J. Knowl. Manag.* **2017**, *21*, 492–515. [CrossRef]
9. Loon, M. Knowledge management practice system: Theorising from an international meta-standard. *J. Bus. Res.* **2017**, 1–10. [CrossRef]
10. Ren, X.; Deng, X.; Liang, L. Knowledge transfer between projects within project-based organizations: The project nature perspective. *J. Knowl. Manag.* **2018**, *22*, 1082–1103. [CrossRef]

11. Rodríguez, G.; Ibarra, M.S.; Cubero, J. Competencias básicas relacionadas con la evaluación. Un estudio sobre la percepción de los estudiantes universitarios. *Educ. XX1* **2018**, *21*, 181–208. [CrossRef]
12. Duderstadt, J.J. Engineering for a changing world. In *Holistic Engineering Education*; Springer: New York, NY, USA, 2010; pp. 17–35, ISBN 978-1441913920.
13. Garcia-Penalvo, F.J. Engineering contributions to a multicultural perspective of the knowledge society. *IEEE Rev. Iberoam. Tecnol. del Aprendiz.* **2015**, *10*, 17–18. [CrossRef]
14. Mohrman, K.; Ma, W.; Baker, D. The research university in transition: The emerging global model. *High. Educ. Policy* **2008**, *21*, 5–27. [CrossRef]
15. Salmi, J. *The Challenge of Establishing World-class Universities*; The World Bank: Washington, DC, USA, 2009; ISBN 978-0821378656.
16. Lavalle, C.; De Nicolás, V.L. Peru and its new challenge in higher education: Towards a research university. *PLoS ONE* **2017**, *12*, 1–12. [CrossRef] [PubMed]
17. Fernández-Pello, C. La vinculación con la sociedad en las universidades de investigación. In *La Universidad: Motor de Transformación de la Sociedad*; Cazorla, A., Stratta, R., Eds.; UPM: Madrid, Spain, 2017; pp. 52–61, ISBN 978-8461794744.
18. Cazorla, A. Hacia una universidad de investigación desde una profesional: Estrategias. In *La Universidad: Motor de Transformación de la Sociedad*; Cazorla, A., Stratta, R., Eds.; UPM: Madrid, Spain, 2017; pp. 16–31, ISBN 978-8461794744.
19. Kristjánsson, K. There is something about Aristotle: The pros and cons of aristotelianism in contemporary moral education. *J. Philos. Educ.* **2014**, *48*, 48–68. [CrossRef]
20. De Nicolás, V.L. Los rankings, un reflejo de la Universidad en el mundo: Webometrics. In *La Universidad: Motor de Transformación de la Sociedad*; Cazorla, A., Stratta, R., Eds.; UPM: Madrid, Spain, 2017; pp. 32–51. ISBN 978-8461794744.
21. Jensen, A.F.; Thuesen, C.; Geraldi, J. The projectification of everything: Projects as a human condition. *Proj. Manag. J.* **2016**, *47*, 21–34. [CrossRef]
22. Hauc, A.; Vrečko, I.; Barilović, Z. A holistic project-knowledge society as a condition for solving global strategic crises. *Drus. Istraz.* **2011**, *20*, 1039–1060. [CrossRef]
23. The Standish Group. *Chaos Report 2015*; The Standish Group: Boston, MA, USA, 2015.
24. Lundin, R.A.; Arvidsson, N.; Brady, T.; Ekstedt, E.; Midler, C.; Sydow, J. *Managing and Working in Project Society. Institutional Challenges of Temporary organizations*; Cambridge University Press: Cambridge, UK, 2015; ISBN 978-1107077652.
25. Crawford, L. Profiling the competent project manager. In Proceedings of the PMI Research Conference: Project Management Research at the Turn of the Millennium, Paris, France, 24 June 2000; pp. 3–15.
26. Morris, P.W.G.; Crawford, L.; Hodgson, D.; Shepherd, M.M.; Thomas, J. Exploring the role of formal bodies of knowledge in defining a profession—The case of project management. *Int. J. Proj. Manag.* **2006**, *24*, 710–721. [CrossRef]
27. Project Management Institute. *PMI's Pulse of the Profession®In-Depth Report: The Competitive Advantage of Effective Talent Management*; PMI: Newtown Square, PA, USA, 2013.
28. Vukomanović, M.; Young, M.; Huynink, S. IPMA ICB 4.0: A global standard for project, programme and portfolio management competences. *Int. J. Proj. Manag.* **2016**, *34*, 1703–1705. [CrossRef]
29. Chipulu, M.; Neoh, J.G.; Ojiako, U.; Williams, T. A multidimensional analysis of project manager competences. *IEEE Trans. Eng. Manag.* **2013**, *60*, 506–517. [CrossRef]
30. Jugdev, K.; Thomas, J.; Delisle, C.L. Rethinking project management - Old truths and new insights. *Int. Proj. Manag. J.* **2001**, *7*, 36–43.
31. International Project Management Association. *Individual Competence Baseline for Project, Programme & Portfolio Management*, 4th ed.; IPMA: Zurich, Switzerland, 2015; ISBN 978-9492338013.
32. International Project Management Association. *Organisational Competence Baseline for Developing Competence in Managing by Projects*, 1st ed.; IPMA: Zurich, Switzerland, 2016; ISBN 978-9492338068.
33. International Project Management Association. *Project Excellence Baseline for Achieving Excellence in Projects and Programmes*, 1st ed.; IPMA: Zurich, Switzerland, 2016; ISBN 978-9492338051.
34. Project Management Institute. *Project Manager Competency Development Framework*, 3rd ed.; PMI: Newtown Square, PA, USA, 2017; ISBN 978-1628250916.

35. Project Management Institute. *Implementing Organizational Project Management: A Practice Guide*; PMI: Newtown Square, PA, USA, 2014; ISBN 978-1628250350.
36. Project Management Institute. *A Guide to the Project Management Body of Knowledge*, 6th ed.; PMI: Newtown Square, PA, USA, 2017; ISBN 978-1628251845.
37. International Organization for Standardization. *ISO 21500: Guidance on Project Management*; ISO: Geneva, Switzerland, 2012.
38. European Union. *The PM2 Project Management Methodology Guide*, 1st ed.; Publications Office of the European Union: Luxembourg, 2016; ISBN 978-9279638725.
39. Takey, S.M.; Carvalho, M.M. Competency mapping in project management: An action research study in an engineering company. *Int. J. Proj. Manag.* **2015**, *33*, 784–796. [CrossRef]
40. Omidvar, G.; Samad, Z.A.; Alias, A. A framework for job-related competencies required for project managers. *Int. J. Res. Manag. Technol.* **2014**, *4*, 1–10.
41. Azim, S.; Gale, A.; Lawlor-Wright, T.; Kirkham, R.; Khan, A.; Alam, M. The importance of soft skills in complex projects. *Int. J. Manag. Proj. Bus.* **2010**, *3*, 387–401. [CrossRef]
42. Ahern, T.; Leavy, B.; Byrne, P.J. Knowledge formation and learning in the management of projects: A problem solving perspective. *Int. J. Proj. Manag.* **2014**, *32*, 1423–1431. [CrossRef]
43. Ojiako, U.; Chipulu, M.; Ashleigh, M.; Williams, T. Project management learning: Key dimensions and saliency from student experiences. *Int. J. Proj. Manag.* **2014**, *32*, 1445–1458. [CrossRef]
44. Ojiako, U.; Chipulu, M.; Marshall, A.; Ashleigh, M.J.; Williams, T. Project management learning: A comparative study between engineering students' experiences in South Africa and the United Kingdom. *Proj. Manag. J.* **2015**, *46*, 47–62. [CrossRef]
45. Cheng, M.I.; Dainty, A.R.J.; Moore, D.R. What makes a good project manager? *Hum. Resour. Manag. J.* **2005**, *15*, 25–37. [CrossRef]
46. Crawford, L. Senior management perceptions of project management competence. *Int. J. Proj. Manag.* **2005**, *23*, 7–16. [CrossRef]
47. Le Deist, F.D.; Winterton, J. What is competence? *Hum. Resour. Dev. Int.* **2005**, *8*, 27–46. [CrossRef]
48. Binkley, M.; Erstad, O.; Herman, J.; Raizen, S.; Ripley, M. *Defining 21st Century Skills. Draft White Paper 1*; The University of Melbourne: Melbourne, Australia, 2010.
49. Onisk, M. *Is Measuring Soft-skills Training Really Possible?* Appcon: Sidney, Australia, 2011.
50. Omidvar, G.; Jaryani, F.; Samad, Z.B.A.; Zafarghandi, S.F.; Nasab, S.S. A proposed framework for project managers' competencies and role of e-portfolio to meet these competencies. *Int. J. e-Education, e-Business, e-Management e-Learning* **2011**, *1*, 311–321. [CrossRef]
51. Teijeiro, M.; Rungo, P.; Freire, M.J. Graduate competencies and employability: The impact of matching firms' needs and personal attainments. *Econ. Educ. Rev.* **2013**, *34*, 286–295. [CrossRef]
52. Cerezo-Narváez, A.; Bastante-Ceca, M.J.; Yagüe-Blanco, J.L. Traceability of intra- and interpersonal skills: From education to labor market. In *Human Capital and Competences in Project Management*; Otero-Mateo, M., Pastor-Fernández, A., Eds.; InTech: Rijeka, Croatia, 2018; pp. 87–110. ISBN 978-9535137870.
53. Bushuyev, S.D.; Wagner, R.F. IPMA Delta and IPMA organisational competence baseline (OCB). *Int. J. Manag. Proj. Bus.* **2014**, *7*, 302–310. [CrossRef]
54. Strang, K. Achieving organizational learning across projects. In Proceedings of the PMI®Global Congress 2003-North America, Baltimore, MD, USA, 25 September 2003; p. 10.
55. Pant, I.; Baroudi, B. Project management education: The human skills imperative. *Int. J. Proj. Manag.* **2008**, *26*, 124–128. [CrossRef]
56. Awan, M.H.; Ahmed, K.; Zulqarnain, W. Impact of project manager's soft leadership skills on project success. *J. Poverty, Invest. Dev.* **2015**, *8*, 27–47.
57. Kandelousi, N.S.; Ooi, J.; Abdollahi, A. Key success factors for managing projects. *Int. Sch. Sci. Res. Innov.* **2011**, *5*, 1185–1189.
58. Cousillas, S.M.; Rodríguez, V.; Villanueva, J.M.; Álvarez, J.V. Análisis de factores de éxito y causas de fracaso en proyectos: Detección de patrones de comportamiento mediante técnicas de clustering. In Proceedings of the 17th International Congress on Project Management and Engineering, Logroño, Spain, 17–19 July 2013; pp. 190–202.
59. Camilleri, E. *Project Sucess: Critical Factors and Behavoiurs*; Gower Publishing: Burlington, VT, USA, 2011; ISBN 978-0566092282.

60. Butler, A.G. Project management: A study in organizational conflict. *Acad. Manag. J.* **2012**, *16*, 84–101.
61. Swanigan, C.L. *Examining the Factors of Leadership in Project Management: A Qualitative Multiple Case Study;* University of Phoenix: Phoenix, AZ, USA, 2016.
62. Kerzner, H. *Project Management Best Practices: Achieving Global Excellence,* 4th ed.; John Wiley & Sons: New Jersey, NJ, USA, 2018; ISBN 978-1119468851.
63. Apenko, S.N.; Klimenko, O.A. Differentiation criteria for strategic and non-strategic project and programmes. In Proceedings of the 12th International Scientific and Technical Conference on Computer Sciences and Information Technologies (CSIT), Lviv, Ukraine, 5–8 September 2017; pp. 52–57.
64. De los Ríos, I.; López, F.; Pérez, C. Promoting professional project management skills in engineering higher education: Project-based learning (PBL) strategy. *Int. J. Eng. Educ.* **2015**, *31*, 184–198.
65. Apenko, S.N.; Romanenko, M.A. Formation of personnel potential of innovative projects based on international professional standards. *Actual Probl. Econ.* **2016**, *186*, 244–252.
66. Slavyanska, V. The project manager's competencies as a basic factor influensing the team effectiveness. *New Knowl. J. Sci.* **2017**, *6*, 17–28.
67. De los Ríos, I.; Cazorla, A.; Díaz-Puente, J.M.; Yagüe, J.L. Project–based learning in engineering higher education: Two decades of teaching competences in real environments. *Procedia Soc. Behav. Sci.* **2010**, *2*, 1368–1378. [CrossRef]
68. Salgado, J.P.; De los Ríos, I.; González, M. Management of entrepreneurship projects from project-based learning: Coworking StartUPS project at Universidad Politécnica Salesiana (Salesian Polytechnic University), Ecuador. In *Case Study of Innovative Projects - Successful Real Cases*; InTech: Rijeka, Croatia, 2017; pp. 227–245.
69. Organisation for Economic Co-operation and Development. *The Definition and Selection of Key Competencies. Executive Summary*; OECD Publishing: Paris, France, 2005; ISBN 978-0889372726.
70. González, J.; Wagenaar, R. *Tuning Educational Structures in Europe. Universities' Contribution to the Bologna Process*; Universidad de Deusto: Bilbao, Spain, 2005; ISBN 978-8498306439.
71. Beneitone, P.; Esquetini, C.; González, J.; Maletá, M.M.; Siufi, G.; Wagenaar, R. *Latin America Project. Tuning Reflections on and Outlook for Higher Education in Latin America*; Universidad de Deusto: Bilbao, Spain, 2007.
72. Guerrero, D.; Palma, M.; La Rosa, G. Developing competences in engineering students. The case of Project Management course. *Procedia Soc. Behav. Sci.* **2014**, *112*, 832–841. [CrossRef]
73. Andersen, N.; Yazdani, S.; Andersen, K. Performance outcomes in engineering design courses. *J. Prof. Issues Eng. Educ. Pract.* **2007**, *133*, 2–8. [CrossRef]
74. Astigarraga, T.; Dow, E.M.; Lara, C.; Prewitt, R.; Ward, M.R. The emerging role of software testing in curricula. In Proceedings of the IEEE Transforming Engineering Education: Creating Interdisciplinary Skills for Complex Global Environments; IEEE: Dublin, Ireland, 2010; pp. 1–26.
75. Kans, M. Applying an innovative educational program for the education of today's engineers. *J. Phys. Conf. Ser.* **2012**, *364*, 012113. [CrossRef]
76. Mulder, K.F. Strategic competences for concrete action towards sustainability: An oxymoron? Engineering education for a sustainable future. *Renew. Sustain. Energy Rev.* **2017**, *68*, 1106–1111. [CrossRef]
77. Palma, M.; Miñán, E.; De los Ríos, I. Generic engineering competences: A comparative study in an international context. In Proceedings of the 15th International Congress on Project Management and Engineering, Huesca, Spain, 6–8 July 2011; pp. 2552–2569.
78. The Accreditation Board of Engineering and Technology ABET. Available online: https://www.abet.org/ (accessed on 30 December 2018).
79. The CDIO Initiative CDIO. Available online: http://www.cdio.org/ (accessed on 30 December 2018).
80. European Network for Accreditation of Engineering Education EUR-ACE. Available online: http://www.enaee.eu/ (accessed on 30 December 2018).
81. Malmqvist, J. A comparison of the CDIO and EUR-ACE quality assurance systems. *Int. J. Qual. Assur. Eng. Technol. Educ.* **2012**, *2*, 9–22. [CrossRef]
82. Zamyatina, O.; Minin, M.; Denchuk, D.; Sadchenko, V. Analysis of engineering invention competencies in standards and programmes of engineering universities. *Procedia Soc. Behav. Sci.* **2015**, *171*, 1088–1096. [CrossRef]
83. Halim, M.H.A.; Buniyamin, N. A comparison between CDIO and EAC engineering education learning outcomes. In Proceedings of the IEEE 8th International Conference on Engineering Education (ICEED), Kuala Lumpur, Malaysia, 7–8 December 2016; pp. 22–27.

84. May, D.; Terkowsky, C. What should they learn?—A short comparison between different areas of competence and accreditation boards' criteria for engineering education. In *Engineering Education 4.0*; Springer: Cham, Switzerland, 2016; pp. 911–921. ISBN 978-3319469157.

85. Truong, T.V.; Ha, B.D.; Le, B.N. CDIO contribution to ABET accreditation. In Proceedings of the 14th International CDIO Conference, Kanazawa Institute of Technology, Kanazawa, Japan, 28 June–2 July 2018.

86. Jolly, A. Program Outcomes and Institutions Management Frameworks as Seen by EUR-ACE and by CTI: A Comparison of Criteria. In *Engineering Education for a Smart Society. GEDC 2016, WEEF 2016. Advances in Intelligent Systems and Computing*; Auer, M., Kim, K.S., Eds.; Springer: Cham, Switzerland, 2018; Volume 627, pp. 121–131, ISBN 978-3319609362.

87. Felder, R.M.; Brent, R. Designing and teaching courses to satisfy the ABET engineering criteria. *J. Eng. Educ.* **2003**, *92*, 7–25. [CrossRef]

88. Passow, H.J. Which ABET competencies do engineering graduates find most important in their work? *J. Eng. Educ.* **2012**, *101*, 95–118. [CrossRef]

89. Bragós, R. Faculty teaching skills in the CDIO initiative. *Rev. Docencia Univ.* **2012**, *10*, 57–73.

90. Crawley, E.F. Creating the CDIO syllabus, a universal template for engineering education. In Proceedings of the Frontiers in Education Conference, Boston, MA, USA, 6–9 November 2002; pp. 8–13.

91. Crawley, E.F.; Lucas, W.A.; Malmqvist, J.; Brodeur, D.R. The CDIO Syllabus v2.0. In Proceedings of the 7th International CDIO Conference, Copenhagen, Denmark, 20–23 June 2011; p. 41.

92. European Network for Accreditation of Engineering Education. *EUR-ACE Framework Standards for the Accreditation of Engineering Programmes*; EUR-ACE: Brussels, Belgium, 2008.

93. Augusti, G. Accreditation of engineering programmes: European perspectives and challenges in a global context. *Eur. J. Eng. Educ.* **2007**, *32*, 273–283. [CrossRef]

94. Augusti, G. EUR-ACE: The European accreditation system of engineering education and its global context. In *Engineering Education Quality Assurance*; Springer: Cham, Switzerland, 2009; pp. 41–49, ISBN 978-1-4419-0554-3.

95. Sangwan, S.; Garg, S. WIL and business graduate skill transfer to workplace. *Horizon* **2017**, *25*, 109–114. [CrossRef]

96. Mitchell, G.W.; Skinner, L.B.; White, B.J. Essential soft skills for success in the twenty first century workforce as perceived by business educators. *Delta Pi Epsil. J.* **2010**, *51*, 43–53.

97. Robles, M.M. Executive perceptions of the top 10 soft skills needed in today's workplace. *Bus. Prof. Commun. Q.* **2015**, *75*, 453–465. [CrossRef]

98. Sutton, N. Why can't we all just get along? *Comput. Canada* **2002**, *28*, 20.

99. Truong, H.T.T.; Laura, R.S.; Shaw, K. New insights for soft skills development in Vietnamese business schools: Defining essential softskills for maximizing graduates' career success. *Int. J. Soc. Behav. Educ. Econ. Bus. Ind. Eng.* **2016**, *10*, 1857–1863.

100. Cleary, M.; Flynn, R.; Thomasson, S. *Employability Skills from Framework to Practice*; Commonwealth of Australia: Melbourne, Australia, 2006.

101. Hodge, K.A.; Lear, J.L. Employment skills for 21st century workplace: The gap between faculty and student perceptions. *J. Career Tech. Educ.* **2011**, *26*, 28–41. [CrossRef]

102. Ramlall, S.; Ramlall, D. The value of soft-skills in the accounting profession: Perspectives of current accounting students. *Adv. Res.* **2014**, *2*, 645–654. [CrossRef]

103. Nusrat, M. Soft Skills for Sustainable Employment: Does it really Matter? *Int. J. Manag. Econ. Invent.* **2018**, *04*, 1835–1937. [CrossRef]

104. Alismail, H.A.; Mcguire, P. 21st century standards and curriculum: Current research and practice. *J. Educ. Pract.* **2015**, *6*, 150–155.

105. Alonso, J.; Sáez, A.; Saraite, L.; Caba, C. The financial of public universities in Spain. In *Financial Sustainability in Public Administration*; Springer: Cham, Switzerland, 2017; Volume 56, pp. 227–254. ISBN 978-3319579610.

106. Di Nauta, P.; Merola, B.; Caputo, F.; Evangelista, F. Reflections on the role of University to face the challenges of knowledge society for the local economic development. *J. Knowl. Econ.* **2018**, *9*, 180–198. [CrossRef]

107. Solarte, L.; Sánchez, L.F. Gerencia de proyectos y estrategia organizacional: El modelo de madurez en gestión de proyectos CP3M© V5.0. *Innovar* **2014**, *24*, 5–18. [CrossRef]

108. Etzkowitz, H. The Entrepreneurial University. In *Encyclopedia of International Higher Education Systems and Institutions*; Shin, J.C., Teixeira, P., Eds.; Springer: Dordrecht, The Netherlands, 2019; pp. 1–5, ISBN 978-9401795531.

109. Antony, J.; Krishan, N.; Cullen, D.; Kumar, M. Lean Six Sigma for higher education institutions (HEIs). *Int. J. Product. Perform. Manag.* **2012**, *61*, 940–948. [CrossRef]

110. Atkinson Alpert, S.; Hartshorne, R. An examination of assistant professors' project management practices. *Int. J. Educ. Manag.* **2013**, *27*, 541–554. [CrossRef]

111. Petersen, S.A.; Heikura, T. Modelling project management and innovation competences for technology enhanced learning. In Proceedings of the International Information Management Corporation, Warsaw, Poland, 27–29 October 2010; pp. 1–9.

112. Hakim, A. Contribution of competence teacher (pedagogical, personality, professional competence and social) on the performance of learning. *Int. J. Eng. Sci.* **2015**, *4*, 1–12.

113. Huljenic, D.; Desic, S.; Matijasevic, M. Project management in research projects. In Proceedings of the 8th International Conference on Telecommunications, Zagreb, Croatia, 15–17 June 2005; pp. 663–669.

114. Hernandez, Y.; Cormican, K. Towards the effective management of social innovation projects: Insights from project management. *Procedia Comput. Sci.* **2016**, *100*, 237–243. [CrossRef]

115. Chin, C.M.; Yap, E.H.; Spowage, A.C. Project management methodology for university-industry collaborative projects. *Rev. Int. Comp. Manag.* **2011**, *12*, 901–918.

116. Golini, R.; Kalchschmidt, M.; Landoni, P. Adoption of project management practices: The impact on international development projects of non-governmental organizations. *Int. J. Proj. Manag.* **2015**, *33*, 650–663. [CrossRef]

117. Bodea, C.-N.; Dascălu, M.-I. Modeling Project Management Competences: An Ontology-Based Solution for Competency-Based Learning. In *Communications in Computer and Information Science*; Lytras, M.D., Ordonez de Pablos, P., Avison, D., Sipior, J., Jin, Q., Leal, W., Uden, L., Thomas, M., Cervai, S., Horner, D., Eds.; Springer: Berlin, Germany, 2010; Volume 73, pp. 503–509, ISBN 978-3642131660.

118. Otero-Mateo, M.; Pastor-Fernández, A.; Portela-Núñez, J.-M. El éxito sostenido desde la perspectiva de la dirección y gestión de proyectos. *DYNA Manag.* **2014**, *2*, 1–9. [CrossRef]

119. Rooij, S.W. van Scaffolding project-based learning with the project management body of knowledge (PMBOK®). *Comput. Educ.* **2009**, *52*, 210–219. [CrossRef]

120. Bayona, S.; Bustamante, J.; Saboya, N. PMBOK as a reference model for academic rResearch management. In *Trends and Advances in Information Systems and Technologies*; Rocha, Á., Adeli, H., Reis, L.P., Costanzo, S., Eds.; Springer: Cham, Switzerland, 2018; Volume 206, pp. 863–876, ISBN 978-3642369803.

121. Linstone, H.A.; Turoff, M. *The Delphi Method: Techniques and Applications*; Turoff, M., Linstone, H.A., Eds.; Addison Wesley: Reading, GB, UK, 2002; ISBN 978-0201042948.

122. Dalkey, N.; Helmer, O. An experimental application of the Delphi method to the use of experts. *Manage. Sci.* **1963**, *9*, 458–467. [CrossRef]

123. Habibi, A.; Sarafrazi, A.; Izadyar, S. Delphi technique theoretical framework in qualitative research. *Int. J. Eng. Sci.* **2014**, *3*, 8–13.

124. Yousuf, M.I. Using experts' opinions through Delphi technique - Practical assessment, research and evaluation. *Pract. Assess. Res. Eval.* **2007**, *12*, 1–8.

125. Afshari, A.R.; Yusuff, R.M.; Derayatifar, A.R. An application of Delphi method for eliciting criteria in personnel selection problem. *Sci. Res. Essays* **2012**, *7*, 2927–2935.

126. Okoli, C.; Pawlowski, S.D. The Delphi method as a research tool: An example, design considerations and applications. *Inf. Manag.* **2004**, *42*, 15–29. [CrossRef]

127. Murry, J.W.; Hammons, J.O. Delphi: A versatile methodology for conducting qualitative research. *Rev. High. Educ.* **1995**, *18*, 423–436. [CrossRef]

128. von der Gracht, H.A. Consensus measurement in Delphi studies. Review and implications for future quality assurance. *Technol. Forecast. Soc. Chang.* **2012**, *79*, 1525–1536. [CrossRef]

129. Ruiz Olabuénaga, J.I. *Técnicas de Triangulación y Control de Calidad en la Investigación Socioeducativa*; Mensajero: Bilbao, Spain, 2003; ISBN 978-8427125698.

130. Hsu, C.-C.; Sandford, B.A. The Delphi technique: Making sense of consensus. *Pract. Assess. Res. Eval.* **2007**, *12*, 1–8.

131. Dalkey, N. *An Experimental Study of Group Opinion*; Rand: Santa Monica, CA, USA, 1969.

132. Shekhar Singh, A. Conducting case study research in non-profit organisations. *Qual. Mark. Res. An Int. J.* **2014**, *17*, 77–84. [CrossRef]

133. Altheide, D.L.; Johnson, J.M. Reflections on interpretive adequacy in qualitative research. In *The SAGE Handbook of Qualitative Research*; Denzin, N.K., Lincoln, Y.S., Eds.; SAGE: Thousand Oaks, CA, USA, 2011; pp. 581–594.

134. Green, R.A. The Delphi technique in educational research. *SAGE Open* **2014**, *4*, 215824401452977. [CrossRef]

135. Altbach, P.G.; Salmi, J. *The Road to Academic Excellence*; The World Bank: Washington, DC, USA, 2011; ISBN 978-0821388051.

136. López-Cámara, A.B.; Eslava-Suanes, M.D.; González-López, I.; González, H.G.; León-Huertas, C. De Skills of university professor and their evaluation. In Proceedings of the Sixth International Conference on Technological Ecosystems for Enhancing Multiculturality, Salamanca, Spain, 24–26 October 2018; ACM Press: New York, NY, USA, 2018; pp. 175–179.

137. Tigelaar, D.E.H.; Dolmans, D.H.J.M.; Wolfhagen, I.H.A.P.; van der Vleuten, C.P.M. The development and validation of a framework for teaching competencies in higher education. *High. Educ.* **2004**, *48*, 253–268. [CrossRef]

138. Kallioinen, O. Defining and comparing generic competences in higher education. *Eur. Educ. Res. J.* **2010**, *9*, 56–68. [CrossRef]

139. Blömeke, S.; Zlatkin-Troitschanskaia, O.; Kuhn, C.; Fege, J. Modeling and measuring competencies in higher education. In *Modeling and Measuring Competencies in Higher Education*; Blömeke, S., Zlatkin-Troitschanskaia, O., Kuhn, C., Fege, J., Eds.; SensePublishers: Rotterdam, The Netherlands, 2013; pp. 1–10, ISBN 978-9460918674.

140. Gregory, J.; Jones, R. "Maintaining competence": A grounded theory typology of approaches to teaching in higher education. *High. Educ.* **2009**, *57*, 769–785. [CrossRef]

141. Heikkilä, A.; Niemivirta, M.; Nieminen, J.; Lonka, K. Interrelations among university students' approaches to learning, regulation of learning, and cognitive and attributional strategies: A person oriented approach. *High. Educ.* **2011**, *61*, 513–529. [CrossRef]

142. Wierschem, D.; Johnston, C. The role of project management in university computing resource departments. *Int. J. Proj. Manag.* **2005**, *23*, 640–649. [CrossRef]

143. Kyte, A. A "Fresh Eyes" look at improving the effectiveness of engineering group design projects. *Eng. Educ.* **2013**, *8*, 81–97. [CrossRef]

144. Bryde, D.; Leighton, D. Improving HEI productivity and performance through project management: Implications from a Benchmarking Case Study. *Educ. Manag. Adm. Leadersh.* **2009**, *37*, 705–721. [CrossRef]

145. Brown, C. Application of the Balanced Scorecard in Higher Education. *Plan. High. Educ.* **2012**, *40*, 40–50.

146. Leentjens, A.F.G.; Levenson, J.L. Ethical issues concerning the recruitment of university students as research subjects. *J. Psychosom. Res.* **2013**, *75*, 394–398. [CrossRef]

147. Lurie, Y.; Mark, S. Professional ethics of software engineers: An ethical framework. *Sci. Eng. Ethics* **2016**, *22*, 417–434. [CrossRef]

148. Uruburu-Colsa, A.; Moreno-Romero, A.; Ortiz-Marcos, I.; Cobo-Benita, J.R. Monitoring communication competence in an innovative context of engineering project management learning. In Proceedings of the IEEE Global Engineering Education Conference (EDUCON), Marrakech, Morocco, 17–20 April 2012; Volume 42, pp. 1–7.

149. Ortiz-Marcos, I.; Uruburu-Colsa, A.; Cobo-Benita, J.R.; Prieto-Remón, T. Strengthening communication skills in an innovative context of engineering project management learning. *Procedia Soc. Behav. Sci.* **2013**, *74*, 233–243. [CrossRef]

150. de Campos, L.C.; Dirani, E.A.T.; Manrique, A.L.; Van Hattum-Janssen, N. *Project Approaches to Learning in Engineering Education. The Practice of Teamwork*; SensePublishers: Rotterdam, The Netherlands, 2012; ISBN 978-9460919589.

151. De los Ríos, I.; Figueroa, B.; Gómez, F. Methodological proposal for teamwork evaluation in the field of project management training. *Procedia Soc. Behav. Sci.* **2012**, *46*, 1664–1672.

152. MacLaren, I. The contradictions of policy and practice: creativity in higher education. *London Rev. Educ.* **2012**, *10*, 159–172. [CrossRef]

153. Kirillov, N.P.; Leontyeva, E.G.; Moiseenko, Y.A. Creativity in engineering education. *Procedia Soc. Behav. Sci.* **2015**, *166*, 360–363. [CrossRef]

154. van Ameijde, J.D.J.; Nelson, P.C.; Billsberry, J.; van Meurs, N. Improving leadership in Higher Education institutions: A distributed perspective. *High. Educ.* **2009**, *58*, 763–779. [CrossRef]

155. Schofield, T. Critical success factors for knowledge transfer collaborations between university and industry. *J. Res. Adm.* **2013**, *44*, 38–56.

© 2019 by the authors. Licensee MDPI, Basel, Switzerland. This article is an open access article distributed under the terms and conditions of the Creative Commons Attribution (CC BY) license (http://creativecommons.org/licenses/by/4.0/).

education sciences

MDPI

Article

Development of Final Projects in Engineering Degrees around an Industry 4.0-Oriented Flexible Manufacturing System: Preliminary Outcomes and Some Initial Considerations

Isaías González *[iD] and Antonio José Calderón[iD]

Department of Electrical Engineering, Electronics and Automation, University of Extremadura,
Avenida de Elvas, s/n, 06006 Badajoz, Spain; ajcalde@unex.es
* Correspondence: igonzp@unex.es; Tel.: +34-924-289-600

Received: 8 October 2018; Accepted: 4 December 2018; Published: 9 December 2018

check for
updates

Abstract: New paradigms such as the Industry 4.0, the Industrial Internet of Things (IIoT), or industrial cyber-physical systems (ICPSs) have been impacting the manufacturing environment in recent years. Nevertheless, these challenging concepts are also being faced from the educational field: Engineering students must acquire knowledge and skills under the view of these frameworks. This paper describes the utilization of an Industry 4.0-oriented flexible manufacturing system (FMS) as an educational tool to develop final projects (FPs) of engineering degrees. A number of scopes are covered by an FMS, such as automation, supervision, instrumentation, communications, and robotics. The utilization of an FMS with educational purposes started in the academic year 2011–2012 and still remains active. Here, the most illustrative FPs are expounded, and successful academic outcomes are reported. In addition, a set of initial considerations based on the experience acquired by the FP tutors is provided.

Keywords: engineering education; flexible manufacturing system; Industry 4.0; final project; automation; supervision; robotics; industrial communications

---

1. Introduction

Manufacturing systems are experiencing the advent of innovative trends brought by the information and communication technologies (ICTs)-enabled digital transformation. New paradigms like Industry 4.0, the Industrial Internet of Things (IIoT), industrial cyber-physical systems (ICPSs), or cloud manufacturing (CM) are the most impactful approaches that have arrived to stay. Indeed, Industry 4.0, the so-called fourth industrial revolution, is based on the wide adoption of the IIoT and ICPSs [1]. The next generation of factories are conceived as smart environments where machines, sensors, and actuators are interconnected to enable collaboration, monitoring, and control [2]. As asserted by Cohen et al. [3], one of the reasons for the fascination of Industry 4.0 is that it is an industrial revolution predicted a priori, which provides various opportunities for companies and research institutes to actively shape the future. The implications of Industry 4.0 and associated technologies reach all types of factories, from small and medium-sized enterprises [4] to broader industries such as chemical manufacturing [5] or oil and gas processing [6].

A number of challenges need to be solved for the real implementation of the Industry 4.0-compliant systems: Networked connection of components, massive data gathering, interoperability handling, a wide adoption of ICTs, high investments, collaborative robotics, cyber-security issues, and enhanced flexibility, just to name a few [7].

Another paramount aspect of Industry 4.0 is its impact on employment. There is an unsolved debate about the increase or decrease of employee numbers as a consequence of Industry 4.0 [8]. What seems to be clear is that diverse job profiles are required, such as automation programmer, robot programmer, informatics specialists, software engineer, data analyst, and cyber security specialist [9]. Only qualified and highly educated employees will be able to control these technologies [9]. In this sense, well-prepared Industry 4.0 engineers need an interdisciplinary understanding of systems, production processes, automation technology, information technology, and business processes [10].

Consequently, education and training play an essential role for successful Industry 4.0 implementation. Contents and skills provided to engineering students must encompass the challenging growth of technological advancement [7]. In other words, universities have to provide students with the ability to manage new trends for effective future professional development [7]. In fact, in engineering degrees this concept is raising an increasing interest from educators, as demonstrated by recent examples where courses scheduling and laboratory designs were conducted taking into account the Industry 4.0 framework [10–13]. Indeed, terms like "engineering education 4.0" or "engineer 4.0" are appearing in the literature [14].

Moreover, not only do higher education institutions have to make efforts in this direction, but industrial manufacturing enterprises have to provide specific training and learning solutions for their employees, especially for engineers, technicians, and operational workers. In this sense, companies must pay attention also to Industry 4.0 education-oriented new approaches.

In general, from an educational perspective, it is mandatory for the utilization of experimental systems to apply theoretical concepts and acquire practical skills. Operation of real devices and tools that future engineers will handle in their professional path constitutes a necessity for students [15,16]. Even more, the higher the level of fidelity and realism, the better learning and training are achieved [17]. Therefore, experimental advanced Industry 4.0-compliant equipment is required to act as a didactic platform for successful learning outcomes. A versatile framework that fulfills the abovementioned requisites is a flexible manufacturing system (FMS). An FMS is a flexible automation facility composed by a set of work stations that perform operations over a product that is being processed. The product is transported using a conveyor or transfer around which stations are deployed, acting as a backbone. A number of control units, sensors and actuators, and supervisory interfaces share information in real time through a communication network. FMS has a modular architecture that allows for reorganizing the work stations for different processes and operations [18]. Other terms are also applied for similar or equal concepts: Flexible manufacturing cell, reconfigurable manufacturing system, reconfigurable assembly system, or modular production system. These types of systems are part of the Industry 4.0 environment due to the fact that they are modular manufacturing systems characterized by integrated sensors and standardized interfaces, which corresponds to the enabling technology of Industry 4.0 [19].

An FMS involves a great amount of technology in an integrated manner. Therefore, Figure 1 is expected to show the versatility of an FMS in being used for Research and Development (R&D) and educational activities. In this sense, diverse scopes can be worked: Supervisory systems, automation, networked communications, instrumentation, production scheduling, interoperability handling, maintenance, robotics, and intelligent control, just to name a few, can be trained through an FMS.

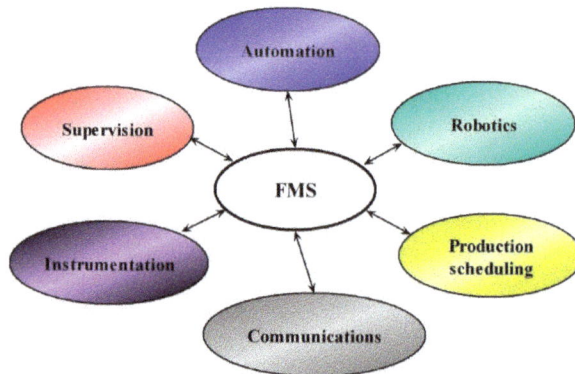

Figure 1. Scopes that can be trained through a flexible manufacturing system (FMS).

In scientific literature, FMSs are profusely used for R&D tasks, for instance in the context of production scheduling [20–23], for integration of the low-cost open source platform Arduino [24], or for applying an Industry 4.0-compliant architecture [25]. An extensive review about applications of FMS can be found in Reference [19].

However, there is a serious scarcity of works dealing with educational approaches that contribute to train the personnel (both engineers and operators) that have to handle such advanced systems. Some recent examples about the utilization of FMS as an educational tool will now be expounded. In the University of León, a FMS has been fully developed and used as a remotely accessible laboratory [26]. Reynard et al. [27] have proposed the educational utilization of a Supervisory Control and Data Acquisition (SCADA) system devoted to an FMS. A didactic FMS and a supervisory system based on LabVIEW were presented in Reference [28]. Various scenarios combining real and virtual processes and controllers by means of an FMS were developed in Reference [29] and were used as an educative environment within a mechatronic laboratory. Toivonen et al. [30] have remarked on the versatility of FMS for engineering education, reporting possible pedagogical applications involving a digital twin of the system. In Reference [31], a group of students performed activities over a full-size didactic FMS in order to evaluate a human–machine cooperation approach toward future manufacturing processes in the Industry 4.0 scenario. Recently, modular manufacturing stations and FMSs have been used as essential parts of Industry 4.0-oriented educational laboratory facilities [10,12]. Learning factory (LF) is also a related concept that in general comprises complex facilities that mimic real production processes and environments, and it is used to develop competencies for present and future personnel [32]. These facilities play a key role in linking academia to industry to spread the culture of innovation [19]. In the present case, the educational-oriented usage of an FMS is simpler, but aims to constitute an initial approximation to such an environment.

The present work reports the utilization of an FMS as a didactic environment to support the development of final projects (FPs) for engineering degrees. An FMS and the most representative FPs are briefly described to provide a perspective of its functionalities and educational capabilities. The FPs were devoted to tasks mainly related to automation, supervision, robotics, and data retrieving. Nonetheless, other important skills can be trained, such as communications, systems integration, sensors and actuators, as will be expounded. Experience accumulated during seven years in the utilization of a real FMS is reported. The successfully achieved academic outcomes are analyzed. The goal of the paper is to provide some early considerations from the perspective of the tutors based on the abovementioned experience.

FMSs are described in the theoretical sessions of courses as a complex and advanced system derived from the evolution of industrial automation equipment. This occurs in a variety of courses, not only in those specifically centered around industrial automation, but in robotics, supervisory systems, and other courses. However, the physical availability of laboratory-scale real industrial FMS equipment is scarce. The main reason is that acquiring an FMS requires a big investment. In fact, the motivation for this paper arose when trying to take advantage of an FMS acquired in the context of an R&D project in collaboration with a manufacturing enterprise. Once the project finished, such valuable equipment was available to be used also for academic activities. Apart from a continuation of the resource utilization, this provided valuable opportunities to teach advanced systems to students, instead of didactic plants. The authors saw the enormous educational potential and decided to offer the opportunity of managing the system under the FP concept. In fact, an FMS is suitable for accommodating different technological approaches as well as pedagogical methodologies.

This paper aims to contribute to fostering a realistic educational environment for automation-related students, mainly oriented to Industry 4.0. Educators can find useful ideas to implement didactic approaches toward an in-depth acquisition of skills for future engineers. Even the information can be used to schedule training sessions for practitioners or workers in manufacturing enterprises.

The remainder of the rest of the paper is as follows. Section 2 deals with organizational aspects of the FPs to give a contextualization of the FMS-based FPs. In Section 3, a description of an FMS is provided jointly with the most demonstrative FPs. In Section 4, the main educational outcomes of the FPs in terms of devoted time, achieved marks, and covered scope are expounded, as well as initial considerations from the educators. Finally, the main conclusions of the work are provided in the last section.

2. FP Procedure

This section aims to provide a brief contextualization of the FPs in the Industrial Engineering School (IES) at the University of Extremadura (UEX) by indicating the main norms and processes involved.

The European Higher Education Area (EHEA) has involved a structural reorganization as well as a shift of the paradigm in universities, emphasizing the relevance of students and the development of skills. Furthermore, at the same time, the global crisis reached academic institutions, reducing the financial resources for acquisition and maintenance of didactic equipment devoted to laboratories.

The IES of the UEX, placed in the city of Badajoz, was founded in 1975, and started the adaptation to the EHEA in the academic year 2009/2010. Nowadays, a total amount of 120 teachers and an average number of 1200 students characterize the school [33].

A paramount part of the engineering degrees is the final project (FP), also called a final year dissertation or a bachelor's or master's thesis, which students have to complete as a final stage prior to achieving their engineer title. The FP consists of an autonomous project developed by the student under the supervision of a tutor. It is intended to reinforce the skills achieved during the degree or to acquire new complementary skills related to their specialization.

The EHEA brought an important novelty regarding FPs, the obligatory development of an FP for every degree. Nonetheless, this was not an authentic novelty for engineering since FPs were already required to finish the degree since its origin. Nowadays, FPs have a duration of 12 European credits, which are equivalent to 300 h, for bachelor's and master's degrees. It is considered to be a course, so it is scheduled in the last semester of the degree. In fact, it has to be presented once the rest of courses have been passed. Note that the bachelor's degree has a duration of four years, whereas the master's degrees last two years. All the norms about FPs and associated documents are publicly available on the webpage of the IES so students can easily find online the information regardless of their particular situation (e.g., academic year, specialization).

The block diagram depicted in Figure 2 shows the stages that must be completed for FP development, from the FP topic definition up to the dissertation event.

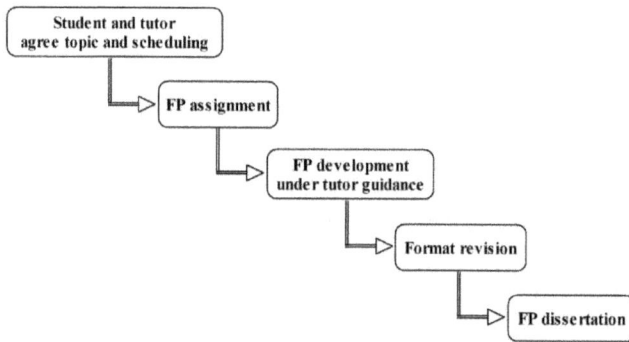

Figure 2. Block diagram of final project (FP) development over time.

As can be observed, the first step consists of selecting or agreeing on the topic. It can be proposed by the student and arranged with the teacher, or delivered from the department offer. Once established, the FP assignment implies the formal compromise of the student, the tutor, and the topic during an academic year.

Each FP can be supervised by one or two tutors belonging to the IES staff. At least one of them must be a teacher of a knowledge area with teaching in the degree of the student. The guidance process of each FP is worth 0.25 European credits for the tutor. Another issue agreed between the tutor and the student consists of the envisioned duration of the FP. Different factors are taken into account such as pending courses and the motivations of the student. In addition, once the project has started, deviations can occur due to a variety of reasons, such as unexpected technical difficulties or learning rhythm variations. In such cases, the tutor establishes a new temporization adapted to the current situation.

The student is supposed to write the project report progressively, during the whole process of the FP. The final document must encompass a series of format rules, so a teacher is designated to revise such issues as a requirement previous to the dissertation. This step is named the "Format revision" in Figure 2.

The dissertation procedure in this university consists of a public event divided into a student-driven exposition and a rigorous argumentation with the tribunal. This tribunal is composed of three professors, responsible for assessing the FP. It must be noted that the tutor does not belong to the assessment tribunal. Figure 3 aims to illustrate the dissertation procedure. Its public nature facilitates the assistance of the student's parents, family, and friends, and also of other FP students aiming to learn about the event. The expected duration is around 30 min.

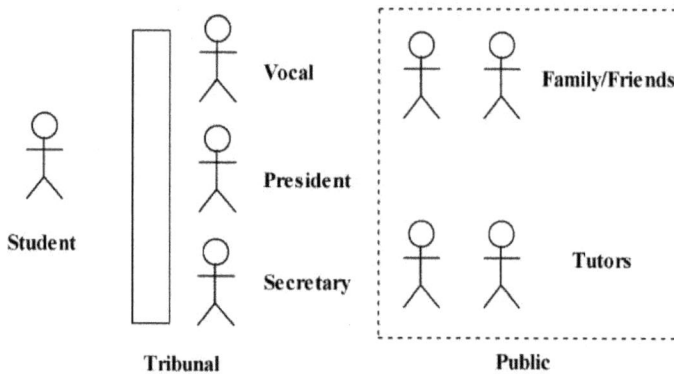

Figure 3. Illustration of the dissertation procedure.

To establish the associated final mark (FM), two rubrics are used by the tribunal in the IES. On the one hand, there is a rubric devoted to assess the project report, report mark (RM), whose weight is 70% of the FM. On the other hand, the dissertation process receives a mark according to another rubric, called the dissertation mark (DM). Consequently, the FM is obtained according to this equation:

$$FM = 0.7 \times RM + 0.3 \times DM. \tag{1}$$

The achievable mark is between 0 and 10, and a distinction award can be reached if the FP receives a mark of 10 and other academic merits have been met. The assessment of this latter consideration is made outside of the tribunal faculties.

For a better comprehension of the students' training in automation and complementary disciplines, Table 1 collects the courses classified into the corresponding degree or master. A noticeable remark is that the Bachelor's Degree in Electronic Engineering and Automation is the most aligned with FMS resources. However, any other degree or master's can be accommodated in such a system. In fact, as aforesaid, one of the advantages and goals of the FPs is to allow students to acquire expertise in fields that have not been deeply studied in previous courses. In the same sense, it must be noted that up to the present date, only FPs for degrees have been carried out, but an FMS is suitable for a PhD thesis.

Table 1. Previous and complementary courses related to automation and supervision.

Title	Course	Character
Bachelor's Degree in Electronic Engineering and Automation	Introduction to Automation	Obligatory
	Automation I	Obligatory
	Automation II	Obligatory
	Industrial Process Control	Obligatory
	Robotics and Perception Systems	Obligatory
	Supervisory Control Systems	Optional
Bachelor's Degree in Electrical Engineering	Introduction to Automation	Obligatory
	Industrial Automation	Obligatory
	Supervisory Systems	Optional
Bachelor's Degree in Mechanical Engineering	Introduction to Automation	Obligatory
	Industrial Automation	Optional
Bachelor's Degree in Industrial Chemical Engineering	Electronic Engineering and Automation	Obligatory
	Process Engineering II	Obligatory
Master's Degree in Industrial Engineering	Electronic Technology and Automation	Obligatory
	Automation of Production Systems	Optional
Master's Degree in Research in Engineering and Architecture	Introduction to Research in Advanced Automation Techniques	Optional

Apart from FPs dealing with FMSs that are expounded in the present paper, the authors also tutor projects about industrial automation and supervision, automated renewable energy systems, and open source devices for data acquisition and automatic control. Some details about the guidance procedure particularly followed by the authors are now commented on. An example of ICT utilization within the presented approach consists of using a Moodle-based learning management system (LMS) for tracking the projects' development. This virtual space is shared between the educators and the students, not only devoted to FMS-based FPs. Tutors upload documents and web links so students can find them easily online. Such information covers aspects related to common procedures about FPs such as norms and advice (about writing the report or the dissertation), as well as meeting requirements. A bidirectional flow of information is established, since both teachers and students can talk through the available forum in the Moodle-LMS, facilitating asynchronous communication and giving students the opportunity for active communication, even among them. Group meetings are carried out mainly in the initial stage of FPs in order to give common information to students regardless of the specific topic. The contents are mainly focused on scheduling aspects of the FPs and on the proposed means of student-tutor communication.

3. FMS and FPs Description

Now that the FP contextualization has been provided, in this section the FMS around which the projects were developed is briefly described. Likewise, in the second subsection, in order to provide an overview of the educational capabilities of an FMS in the Industry 4.0-related field, the most illustrative FPs are explained.

3.1. FMS Description

The utilized FMS is manufactured by the company SMC (Tokyo, Japan) [34]. It accomplishes the assembly and storage of a turning mechanism composed of the following elements: Body, bearing, shaft, cap, and screws. The set is transported in a pallet over the conveyor belt or transfer that acts as a backbone linking the different stations. To complete the assembly, the whole system consists of eight stations. However, currently the FMS is implemented with four stations, those numbered as 1, 7, and 8, and an empty station that is being equipped. Each station has a table-like structure where the components are mounted, such as robots, pneumatic cylinders and distributors, motors, and sensors [18]. In addition, each station is automated by a programmable logic controller (PLC), and all of them are integrated into a fieldbus Process Fieldbus (PROFIBUS) in order to exchange operational information such as sensor signals or control commands.

Concerning the function of the stations, the so-called station 1 (S1) supplies the body and checks its correct position. The next station is the number 7 (S7), which consists of a robotized screwing through an industrial robot arm of the manufacturer ABB (Zürich, Switzerland). Station 8 (S8) implements an automatic warehouse by means of a 2D Cartesian robot. As aforementioned, the last station is nowadays being equipped, namely with a 3D Cartesian robot. Finally, a modular transfer composed of conveyor belts enables the joint operation of all the stations. Over this transfer, the pallets transport the set of pieces between stations. A block diagram of the FMS is depicted in Figure 4 to portray the layout of the stations. As can be observed, the four stations are deployed to achieve a closed square circuit, allowing a continuous and cyclic execution.

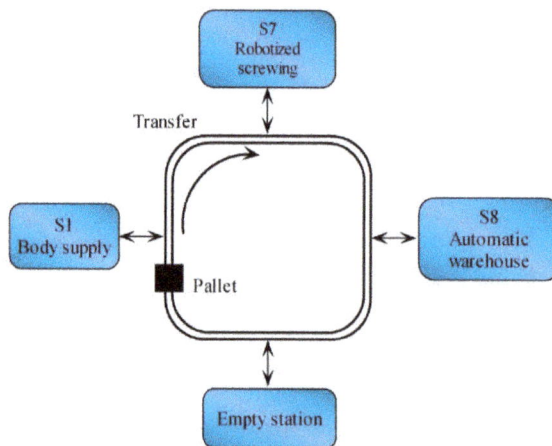

Figure 4. Block diagram of the FMS showing the deployment of the stations.

An assembled turning mechanism mounted over a transport pallet is appreciable in Figure 5a. The whole FMS in the laboratory is shown in Figure 5b. Further details about the FMS can be found in Reference [18].

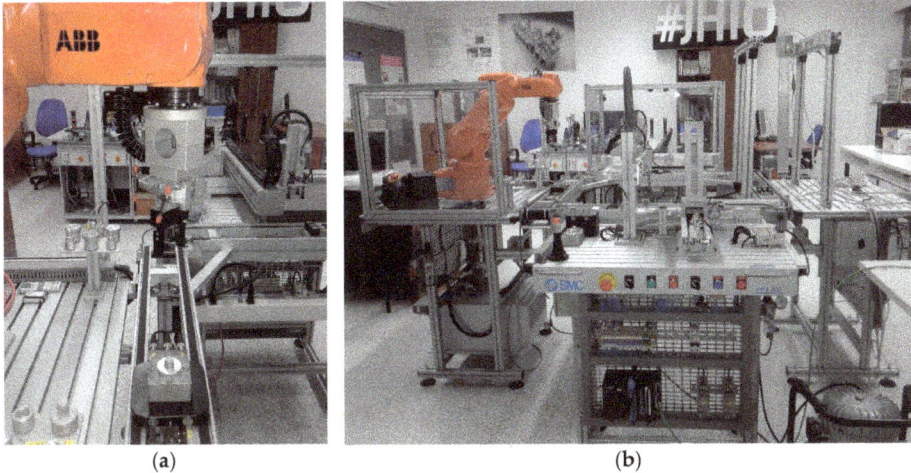

Figure 5. Aspects of the experimental components: (**a**) Assembled turning mechanism over a pallet; (**b**) complete FMS.

Apart from the description of the FMS, at this point it is appropriate to explain diverse security considerations that have been taken into account during the projects' development aiming to reduce potential risks both for students and for the equipment. Damage can be caused by physical reasons, unwanted users, or electrical discharges. Table 2 enumerates the security means that have been implemented classified according to the type of risk.

Table 2. Risks and the associated security means.

Risk	Security Means
Electrical	Electrical protections against discharges
Physical	Physical barriers surrounding the FMS; emergency buttons; limited access to the laboratory
Software	Boundaries in programmable logic controller (PLC) code; user authentication for supervisory management
Cyber threats	Isolation of the local area network (LAN) devoted to FMS-related communications

3.2. Illustrative FPs

Among the 12 FPs that have been carried out using the FMS, four of them are now concisely described to illustrate the scope that has been fruitfully covered.

3.2.1. First-Step Project: Automation of the First Version of the FMS

In this FP, the different stations that composed the initial FMS were integrated from a logical point of view. On the one hand, the fieldbus PROFIBUS was used to interconnect the PLCs of the stations as well as sensors and actuators, and on the other hand, the coordination of the automatic operation was programmed in those PLCs to achieve proper behavior of the FMS. The software STEP7 of the TIA Portal environment was utilized to configure and parameterize the system. The definition of the automation network architecture in such software is seen in Figure 6. It must be noted that in the initial stage of the FMS, only three stations were available. The fourth station (the empty one) was acquired some years later and added to the FMS through another FP.

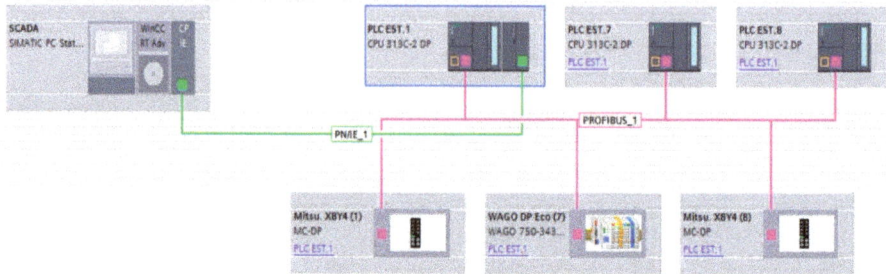

Figure 6. Definition of the automation network architecture for the FMS.

3.2.2. Supervisory Control and Data Acquisition (SCADA) System Development

In this FP, a supervisory system was designed and implemented in order to monitor and manipulate all the elements that composed the FMS. The package WinCC of the suite TIA Portal of Siemens was used for this purpose. The runtime application runs in a PC that is connected to the master PLC of the FMS with the goal of exchanging all the data. Such a connection was implemented via Industrial Ethernet (IE) in order to meet the standard means of communication in Industry 4.0 [18]. An intuitive and user-friendly interface was designed in order to illustrate the status and behavior of the FMS. Two sample screens of the SCADA system are shown in Figure 6. The interface devoted to visualizing the operation of the S7 is seen in Figure 7a, and the screen for monitoring the S8 corresponds to Figure 7b. It should be noted that this FP and the previous one were developed by two students simultaneously and coordinately in order to accomplish the FMS in an effective manner.

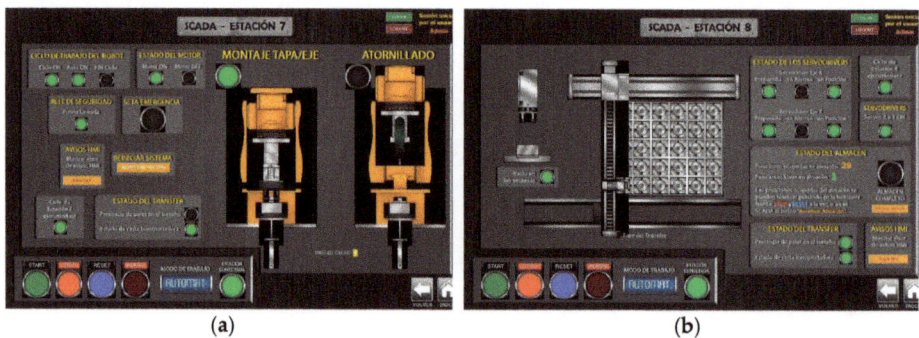

Figure 7. Screens of the supervisory control and data acquisition (SCADA) system developed in a FP: (a) Screen to monitor the operation of the S7; (b) screen to monitor the operation of the S8.

3.2.3. Data Storage in a Database

Within industrial utilization of manufacturing systems, the proper storage and treatment of information plays a vital role both for an effective operation and for higher hierarchical levels that manage software applications such as enterprise resource planning (ERP) or manufacturing execution systems (MESes). Under this perspective, in this FP a data storage mechanism was implemented for the signals handled in the FMS. Structured query language (SQL) reports were generated to record information in an in-house database, which is an accessible data source for other applications. Within the SCADA system, a set of scripts coded in Visual Basic (VB) are responsible for launching the data recording task. In Figure 8a the deployment of the database is schematized through a block diagram, whereas in Figure 8b a part of the designed VB code is shown.

```
Sub NuevoRegistroSQL()
'   declaración de variables locales
Dim conn, rst, SQL_Table

On Error Resume Next

Set conn = CreateObject("ADODB.Connection")
Set rst = CreateObject("ADODB.Recordset")

'abrir el origen de datos de datos SQL
conn.Open "Provider=MSDASQL;Initial Catalog=FMSdb; DSN=FMSdatabase"

'rutina de error al realizar la conexión con la base la datos
If Err.Number <> 0 Then
    ShowSystemAlarm "Error #" & Err.Number & " " & Err.Description
    'muestra alarma del error, su número y una descripción
    Err.Clear
    Set conn = Nothing
    Exit Sub
End If
```

(b)

(a)

Figure 8. Deployment of a database for the FMS: (**a**) Block diagram of the approach; (**b**) details of the code of a Visual Basic (VB) script.

3.2.4. Anthropomorphic Robot Arm Programming

The programming of a robotic arm was carried out in this project. The trajectory that the robot performed was defined in robotics application programming interactive dialogue (RAPID) language, so the students could modify it in order to execute different tasks. For instance, one of the FPs devoted to robotics consisted of programming the tracking of a triangular surface and also the drawing of some geometric shapes. ABB Robot Studio software was used both for the simulation and definition of the new trajectories. In Figure 9a, an image of the 3D representation of a robotic arm following a triangular surface can be observed. A star drawing traced by the robot can be seen in Figure 9b.

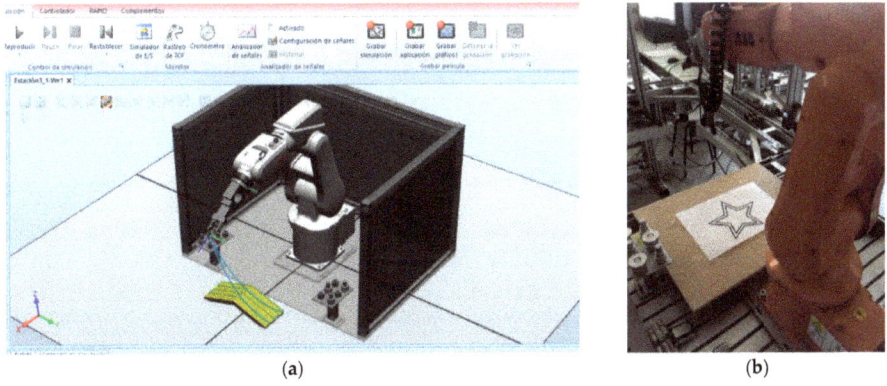

(a)   **(b)**

Figure 9. Aspect of the tasks developed in a robotics-devoted FP: (**a**) Screenshot of the ABB Robot Studio software used to program a surface tracking application; (**b**) photograph of a drawing performed by the robotic arm.

Apart from utilization in the FP context, the FMS constitutes a useful resource for different didactic tasks. For instance, in order to attract future students, the university organizes guided visits for high school students during the month of March. Within the engineering laboratories, the FMS is shown to those students, since it provides attractive insights into the spheres of automation and robotics. Indeed, the FMS performs a working cycle to demonstrate the integration of the stations, since the anthropomorphic robot is the most eye-catching element according to the impressions expressed by the students. In a similar way, this system is presented to students of optional courses devoted to supervisory systems as an example of a real and advanced environment where the concepts and skills that they are learning can be applied.

4. Discussion

The development of the above-described FPs has given rise to a powerful benchmark where students, and even also teachers, can learn about advanced automation, supervision, systems integration, industrial network communications, sensors and actuators, robotics, and so forth. This multidisciplinary skills support is a key feature of the FMS, which makes it suitable to be utilized as an experimental benchmark under the Industry 4.0 focus. The discussion conducted in this section is divided into three parts. The first one is dedicated to describing the educational outcomes of the FPs through these parameters: Devoted time, achieved mark, and covered scope. The second subsection reports the considerations from the tutors' perspectives, covering both advantages and disadvantages. Finally, emerging trends in manufacturing and automation that can be trained in the FMS are commented on.

4.1. Academic Outcomes

The obtained marks have been very good in every FP, as can be observed in Figure 10, where the number of developed FPs and their corresponding mean marks and devoted months from academic year 2011–12 up to the present date are shown. It should be clarified that the spent time was considered from the real start of the project, not from the assignment document.

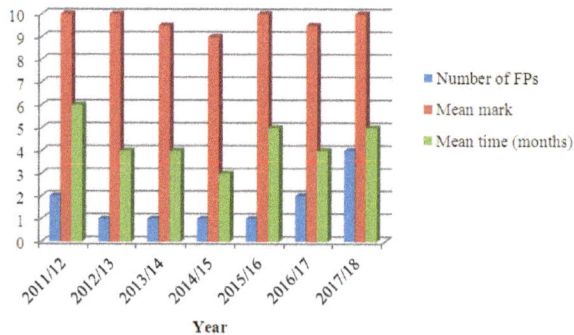

Figure 10. Number of developed FPs and their corresponding mean marks and devoted months.

The mean time devoted to developing the projects was around 4.5 months. The FP course FP is supposed to last a semester (i.e., 12 credits), as explained in the second section. However, as it is well known, the FP is commonly performed in the last stage of the degree. Therefore, the devoted time is very variable depending on the particular situation and motivation of the students.

Concerning the number of FPs, it must be noted that in the recently finished academic year, four projects were performed, which is the largest amount to date. At the present moment, two FPs are being started in the FMS, but they have not been considered for the manuscript.

The development of FPs supported by the FMS is feasible according to the described academic outcomes. This advanced facility has promoted the introduction of the Industry 4.0 paradigm in engineering degrees. In addition, apart from the four reported projects, another eight have been carried out using the FMS. Most of them were devoted to programming the Cartesian and anthropomorphic robots. For instance, the last two FPs were devoted to designing and automating a 3D Cartesian robot in the empty station. Figure 11 illustrates the classification of the FPs according to their main scope. In other words, every project has involved all the scopes at a higher or lower level, but in order to provide insight into the capabilities of the FMS, they have been classified taking into account the principal goal. As can be observed, those devoted to automation and robotics were prevalent, four FPs for each scope. This is coherent with the fact that most students have passed a course entitled "Automation II" where both topics are the core of this course.

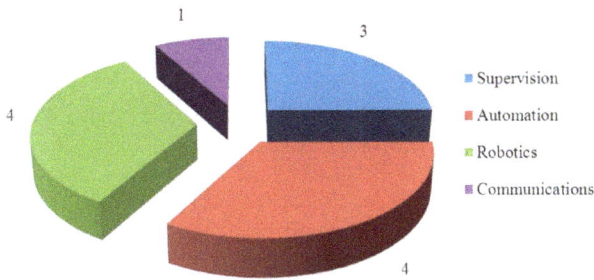

Figure 11. Classification of developed FPs regarding their main scope.

With the goal of illustrating the different elements with which students have been training, Table 3 enumerates such elements divided into groups depending on their nature. Special attention must be paid to the fact that, as is evident, all the handled entities were experimentally studied in an integrated way, not in isolated form. For instance, the measurements provided by sensors were processed and transmitted via fieldbus to an automation unit and a supervisory interface. The student is expected to experience a complete comprehension of the joint operation of the whole system, according to the Industry 4.0 concept.

Table 3. Elements handled by students during FPs using the FMS.

Software	Sensors	Actuators	Control Units	Supervisory Devices	Network Fieldbus
TIA Portal SQL Manager Robot Studio	Presence Position Encoder	Electrical Pneumatic Servomotor	PLC S7-300 and S7-1200 Servomotor driver Robot controller	Windows PC Runtime KTP 600	PROFINET/Ethernet PROFIBUS

4.2. Considerations from Tutors

Through a set of semistructured interviews, the opinions of the tutors were collected using a qualitative approach. The interviews took place once the FPs were completely finished, with the goal of registering the impressions of the tutors about the development and outcomes of the FPs. The semistructured interviews involved three open-ended questions to identify the advantages and drawbacks of using the FMS for FPs, as well as improvements that the educator planned to introduce in future projects. The most descriptive comments and reflections from tutors are shown in Table 4.

Table 4. Descriptive comments and reflections of tutors gathered through semistructured interviews.

Advantages	Drawbacks
To my knowledge, this is the first time that the students have managed equipment with a focus on the Industry 4.0. They have heard about the concept, the new industrial revolution, but they have not really handled any laboratory system compliant with it. This is imperative for them to start to apprehend Industry 4.0 implications.	Using proprietary software packages imposes an important limit for the student because he/she cannot follow programming or configuring tasks at home: Their presence in the laboratory is imposed. For the next projects, the utilization of open source software and hardware should be approached.
Each new project relies to a high degree on the previous ones. For instance, the inclusion of the new station was based on the automation and communication of the rest of the stations. This is seen as a positive feature since students do not learn isolated technologies but an integrated and multidisciplinary environment.	Tutoring a project in the Industry 4.0 is an exciting task: However, the devoted time and effort have been higher than in the case of covering automation systems with more traditional focuses. I have dedicated a lot of time to studying the state of the art as well as the Industry 4.0 requirements.

As a consequence of these collected perceptions, a set of initial considerations about the process of performing FPs around the FMS are expounded hereafter. To begin with, the main benefits found by the educators are now commented on:

- The handling of real industrial equipment constitutes one of the most significant advantages of the system. The FMS is composed of industrial components and is not specifically designed for educational purposes. Students learn and train using the same equipment and technologies that they will face in future professional environments. From the technical perspective, this feature facilitates the modification, addition, or reparation of components, since they use standardized signal ranges, communication protocols, and so forth.
- Accommodation of a number of emerging trends: The FMS can integrate different innovative approaches such as open source tools, virtual environments, and condition monitoring. This will be thoroughly commented on in the next subsection.
- The FMS does not suppose a disruptive scenario, since it is composed of well-known equipment such as PLCs and SCADA systems, as well as sensors and actuators. This feature contributes to facilitate the learning process of the educators, who in fact are also required to acquire deep expertise on innovative paradigms.
- Another benefit is the transversality of the FMS derived from the coexistence of different technologies common to engineering training such as ICT, informatics, and hardware skills, thus enabling the development of multidisciplinary projects.
- Applicability of the FMS for different degrees: As it has been commented on, the degree in Electronic Engineering and Automation is the most appropriate to deal with the FMS. However, other specializations that can also take advantage of the FMS capabilities are Electrical Engineering, Mechanical Engineering, and Chemical Engineering. In addition, out of the typically industrial engineering scope, other disciplines can be supported such as those devoted to telecommunications and informatics. They can be trained around the FMS with an orientation toward advanced networked communications and cyber-security issues, directly related to the Industry 4.0.
- Students become familiar with advanced concepts related to Industry 4.0, acquiring the basis to successfully develop their professional tasks within this challenging framework.
- Students are free to design the solutions that the FP implements, so their creative thinking and curiosity are expected to be stimulated. That is, the tutors guide the development but do not impose boundaries upon the incorporation of the ideas of the students. For instance, when designing the SCADA system, the student decided the layout, visual aspects, and other items according to his own criteria.
- All the FPs were agreed on by students and tutors instead of chosen from a list of available topics. This is considered to be a signal of the attraction that the FMS generates among the students.
- Different levels of autonomy were detected by the tutors. One of the main goals of the FPs is to promote the autonomy of students to learn by themselves and to solve troubles found in their path.
- Operational staff training could also be performed, for instance within a collaboration framework with industrial enterprises or vocational educational institutions. The integration of workers into an Industry 4.0 system is signaled as a significant challenge [35]: Even the term "operator 4.0" is gaining attention in the literature [35,36].

As drawbacks, the following ones have been found:

- Utilization of proprietary technology both for software and hardware tools: For instance, this fact implies high costs in the acquisition of software licenses. It must be noted that, in order to solve this issue, some effort is being currently devoted to using open source hardware and software in the FMS. In addition, as is evident, the students acquire necessary knowledge and skills regardless of the particular software or hardware resources.

- Previous expertise is required from students before the assignment of the FP. Aiming at maximizing the learning outcomes of the students, a minimum level of previous knowledge and skills are desirable before working with the FMS. The FMS is a complex and valuable equipment, so nowadays the criteria of the teachers is that students must have passed previous courses, mainly dealing with automation. Nonetheless, as it was stated in the second section, the FP offers an opportunity to reinforce or to acquire new knowledge and skills.
- Impossibility of performing a number of tasks outside of the physical laboratory: Students have to physically use the FMS for most of the FPs, so they need access to the laboratory and to stay inside during the required time. In other words, scarce tasks can be performed out of the laboratory, and therefore the FMS-based FPs are not suitable for those students that have some reason not to attend the laboratory (e.g., job, illness, childcare, or elder care).
- The tutoring task is scarcely considered within the teaching occupation. Despite the fact that tutors are commonly proud of encouraging the academic growth of FP students, the time and effort dedicated are not correctly valorized. The guidance of a FP is a time-consuming task, especially if it involves an experimental complex facility like the FMS. As was indicated in the second section, tutoring a FP is worth 0.25 European credits for the teacher. This is especially important when educators must face challenging topics such as the Industry 4.0 and related paradigms.
- Utilization of the FMS under project-based learning (PBL) methodology: Out of the FPs, within automation-related courses the FMS could be used to develop projects following the PBL technique so students have an active role and face a number of tasks that reinforce the contents provided during classes. Despite being an interesting application, deep expertise is required in order to avoid damage to the components of the FMS and of the students, though for advanced courses, a PBL approach could be conducted under the proper supervision of an educator.
- A small number of developed FPs around the FMS: Despite the fact that FPs have been successfully completed, the number of projects, and hence the number of involved students, is still small. In an exercise of self-criticism, this is considered to be a limitation of the presented work. Further analyses about pedagogical methodologies and the acquisition of Industry 4.0-related skills will be conducted when a larger amount is reached.
- Higher introduction of the Industry 4.0 paradigm in the engineering degrees: The challenges that this complex concept involves for educational institutions are still being discovered. It does not only require managing advanced equipment (hardware and software), but it implies a framework where a number of disciplines are orchestrated in an integrated manner. In this sense, the FMS has demonstrated itself to be a powerful platform in covering a number of scopes and is expected to incorporate new Industry 4.0-compliant features. Apart from this, institutions, particularly at the university level, need to update and adapt theoretical and practical content toward this new reality. This is in reference to both curricula and available equipment. Educators are also engaged in this process, since they have to modernize their own knowledge and skills. Even more, social aspects of the real deployment of Industry 4.0 must be tackled also in engineering degrees in order to generate successful engineers.

4.3. Modern Manufacturing Trends Related to FMS

Now that the developed FPs and associated outcomes have been expounded and discussed, the goal of this subsection is to provide an overview of the possibilities of the FMS in serving as a benchmark in a number of technological innovative trends. These possibilities highlight the versatility of the FMS and promote its utilization for a number of new FPs dealing with advanced concepts. This is a very valuable feature for didactic and educative systems. Some of the most recent trends in manufacturing and automation systems in the Industry 4.0 arena as well as their inclusion in the FMS via FPs are commented on thereafter.

- Sustainability and environmental concerns achieve increasing attention for manufacturing systems [37]: Hence, various approaches can be carried out in the FMS in that sense. The monitoring of energy consumption in order to be considered in production planning or scheduling is an example.

- Condition monitoring and data acquisition are directly linked to advanced maintenance techniques such as prognostics and health management (PHM) schemes. The widespread deployment of various types of sensors makes it possible to achieve so-called smart monitoring [38]. In this sense, sensors to measure temperature and vibrations of the motors of the conveyor and the robotic arm will be added to provide information that will be used for PHM analysis. In a similar way, data-driven models and algorithms to improve decision-making processes require massive data acquisition and are aligned with big data analytics. These issues can also be handled using the FMS as a benchmark to deploy a number of sensors and process the gathered data.

- Cyber-security is a paramount issue in modern automation networks, so its efficient management must be performed by engineering students [39]. Protections in the SCADA system will be the main issue to handle in the FMS.

- Open source hardware and software resources are receiving efforts in the automation and monitoring scopes [40–42]. Indeed, these resources are being signaled as key accelerators for the industry adoption of the IoT [43]. Supervisory systems and microcontrollers of this nature can be accommodated in the FMS through FPs to evaluate their functionalities. Even more, students can use their own devices, which constitutes a modern trend termed bring your own device (BYOD), aiming to encourage their motivation. Particularly, inexpensive open source hardware platforms such as the Arduino microcontroller or the Raspberry Pi microprocessor can be applied under such a movement.

- Online remote laboratories constitute a useful resource for educational purposes [44] and are also under the umbrella of the Industry 4.0 [45]. Indeed, the above-mentioned open source technology facilitates the deployment of remote laboratories [46], so their application to the FMS could promote its utilization as an online laboratory.

- Virtualization of manufacturing processes through augmented or virtual reality is an emergent trend [37,47,48]. For instance, the design of a 3D virtual world to provide immersive experiences within the automated processes will enrich the motivation and skills of students.

- Radio frequency identification (RFID) technology is a key technology for Industry 4.0 and IIoT scenarios [49]. Its integration, through proprietary or open source means, in the FMS will enable traceability and continuous tracking purposes.

- Additive manufacturing or 3D printing is considered to be one of the enabling technologies of Industry 4.0 [19]. The 3D Cartesian robot that is under development will allow for introducing this technology in the FMS.

- Other developments such as those related to CM possibilities [50] or advanced human-machine interactions [31] can also be accommodated in the FMS.

5. Conclusions

This paper presented the utilization of an industrial experimental FMS as a didactic platform in the context of Industry 4.0, enabling the development of FPs in engineering degrees. The utilization of the FMS for this purpose started in the academic year 2011–2012 and still remains active, with a number of FPs being carried out nowadays. A total number of 12 projects have concluded up to the present date. As a sample, four of them were expounded, namely those related to automation, robotics, supervisory systems, and data management tasks.

The academic outcomes achieved by the students illustrate a fruitful utilization of the FMS. A set of initial considerations from the tutors of the FPs were also provided in order to offer useful insights for other educators and researchers.

The FMS provides a versatile environment since it integrates diverse technologies as well as hardware and software entities among which students can train a multiplicity of skills aligned with Industry 4.0 principles, empowering the education of students in such a challenging modern paradigm. Nevertheless, there is still a long way to go for evaluating the benefits of using the FMS as well as the acquisition of Industry 4.0-related skills through a larger amount of FPs.

Future guidelines include the development of FPs as well as PhD theses considering the innovative trends commented on in Section 4.3, those being the open source technology integration currently approached. Moreover, a survey to collect the students' opinions is being designed and applied. Additionally, in-depth qualitative analysis will be performed in further works. The results will be analyzed and will serve as constructive feedback about the learning experience around the FMS.

Author Contributions: Conceptualization, I.G. and A.J.C.; Investigation, I.G. and A.J.C.; Methodology, I.G. and A.J.C.; Resources, I.G. and A.J.C.; Validation, I.G. and A.J.C.; Writing-Original Draft Preparation, I.G. and A.J.C.; Writing-Review & Editing, I.G. and A.J.C.

Funding: This research received no external funding.

Acknowledgments: The authors are grateful to the students who developed the projects under their tutoring. In addition, the authors wish to thank the anonymous reviewers for their valuable suggestions that allowed for improving this article.

Conflicts of Interest: The authors declare no conflict of interest.

Abbreviations

The following abbreviations were used in this manuscript:

BYOD	Bring your own device
CM	Cloud manufacturing
DM	Dissertation mark
EHEA	European Higher Education Area
ERP	Enterprise resource planning
ICPS	Industrial cyber-physical systems
ICT	Information and communication technology
IE	Industrial Ethernet
IES	Industrial Engineering School
IIoT	Industrial Internet of Things
FM	Final mark
FMS	Flexible manufacturing system
FP	Final project
LAN	Local area network
LF	Learning factory
LMS	Learning management system
MES	Manufacturing execution systems
PBL	Project-based learning
PHM	Prognostics and health management
PLC	Programmable logic controller
PROFIBUS	Process Fieldbus
RAPID	Robotics application programming interactive dialogue
RFID	Radio frequency identification
RM	Report mark
SCADA	Supervisory control and data acquisition
SQL	Structured Query Language
UEX	University of Extremadura
VB	Visual Basic

Educ. Sci. **2018**, *8*, 214

References

1. Ismail, A.; Kastner, W. A middleware architecture for vertical integration. In Proceedings of the 1st International Workshop on Cyber Physical Production Systems (CPPS), Vienna, Austria, 12 April 2016. [CrossRef]

2. Iglesias-Urkia, M.; Orive, A.; Barcelo, M.; Moran, A.; Bilbao, J.; Urbieta, A. Towards a lightweight protocol for Industry 4.0: An implementation based benchmark. In Proceedings of the IEEE International Workshop of Electronics, Control, Measurement, Signals and Their Application to Mechatronics, San Sebastian, Spain, 24–26 May 2017. [CrossRef]

3. Cohen, Y.; Faccio, M.; Gabriele, F.; Mora, C.; Pilati, F. Assembly system configuration through Industry 4.0 principles: The expected change in the actual paradigms. *IFAC-PapersOnLine* **2017**, *50*, 14958–14963. [CrossRef]

4. Müller, J.M.; Voigt, K.-I. Sustainable Industrial Value Creation in SMEs: A Comparison between Industry 4.0 and Made in China 2025. *Int. J. Precis. Eng. Manuf.-Green Technol.* **2018**, *5*, 659–670. [CrossRef]

5. Ji, X.; He, G.; Xu, J.; Guo, Y. Study on the mode of intelligent chemical industry based on cyber-physical system and its implementation. *Adv. Eng. Softw.* **2016**, *99*, 18–26. [CrossRef]

6. Redutskiy, Y. Conceptualization of smart solutions in oil and gas industry. *Procedia Comput. Sci.* **2017**, *109*, 745–753. [CrossRef]

7. González, I.; Calderón, A.J.; Figueiredo, J.; Sousa, J.M.C. Design of an educational platform for automation and supervision under the Industry 4.0 framework. In Proceedings of the 12th International Technology, Education and Development Conference (INTED), Valencia, Spain, 5–7 March 2018.

8. Müller, J.M.; Kiel, D.; Voigt, K.-I. What Drives the Implementation of Industry 4.0? The Role of Opportunities and Challenges in the Context of Sustainability. *Sustainability* **2018**, *10*, 247. [CrossRef]

9. Benešová, A.; Tupa, J. Requirements for Education and Qualification of People in Industry 4.0. *Procedia Manuf.* **2017**, *11*, 2195–2202. [CrossRef]

10. Simons, S.; Abé, P.; Neser, S. Learning in the AutFab—The fully automated Industrie 4.0 learning factory of the University of Applied Sciences Darmstadt. *Procedia Manuf.* **2017**, *9*, 81–88. [CrossRef]

11. Prinz, C.; Morlock, F.; Freith, S.; Kreggenfeld, N.; Kreimeier, D.; Kuhlenkötter, B. Learning Factory modules for smart factories in Industrie 4.0. *Procedia CIRP* **2016**, *54*, 113–118. [CrossRef]

12. Madsen, O.; Moller, C. The AAU Smart Production Laboratory for teaching and research in emerging digital manufacturing technologies. *Procedia Manuf.* **2017**, *9*, 106–112. [CrossRef]

13. Mourtzis, D.; Vlachou, E.; Dimitrakopoulos, G.; Zogopoulos, V. Cyber- Physical Systems and Education 4.0—The Teaching Factory 4.0 Concept. *Procedia Manuf.* **2018**, *23*, 129–134. [CrossRef]

14. Ramirez-Mendoza, R.A.; Morales-Menendez, R.; Iqbal, H.; Parra-Saldivar, R. Engineering Education 4.0—Proposal for a new Curricula. In Proceedings of the IEEE Global Engineering Education Conference (EDUCON), Canary Islands, Spain, 17–20 April 2018. [CrossRef]

15. Rojko, A.; Hercog, D.; Jezernik, K. E-training in mechatronics using innovative remote laboratory. *Math. Comput. Simul.* **2011**, *82*, 508–516. [CrossRef]

16. Villa-López, F.H.; García-Guzmán, J.; Vélez Enríquez, J.; Leal-Ortíz, S.; Ramírez-Ramírez, A. Electropneumatic system for industrial automation: A remote experiment within a web-based learning environment. *Procedia Technol.* **2013**, *7*, 198–207. [CrossRef]

17. Balid, W.; Abdulwahed, M. The Impact of Different Pre-Lab Preparation Modes on Embedded Systems Hands-on Lab. In Proceedings of the 9th Annual American Society for Engineering Education (ASEE) Global Colloquium on Engineering Education, Budapest, Hungary, 12–15 October 2009.

18. Calderón Godoy, A.J.; González Pérez, I. Integration of Sensor and Actuator Networks and the SCADA System to Promote the Migration of the Legacy Flexible Manufacturing System towards the Industry 4.0 Concept. *J. Sens. Actuator Netw.* **2018**, *7*, 23. [CrossRef]

19. Bortolini, M.; Galizia, F.G.; Mora, C. Reconfigurable manufacturing systems: Literature review and research trend. *J. Manuf. Syst.* **2018**, *49*, 93–106. [CrossRef]

20. Girbea, A.; Suciu, C.; Nechifor, S.; Sisak, F. Design and implementation of a service-oriented architecture for the optimization of industrial applications. *IEEE Trans. Ind. Inform.* **2014**, *10*, 185–196. [CrossRef]

21. Kammoun, M.A.; Ezzeddine, W.; Rezg, N.; Achour, Z. FMS Scheduling under Availability Constraint with Supervisor Based on Timed Petri Nets. *Appl. Sci.* **2017**, *7*, 399. [CrossRef]

22. Scholze, S.; Barata, J.; Stokic, D. Holistic Context-Sensitivity for Run-Time Optimization of Flexible Manufacturing Systems. *Sensors* **2017**, *17*, 455. [CrossRef]
23. Priore, P.; Ponte, B.; Puente, J.; Gómez, A. Learning-based scheduling of flexible manufacturing systems using ensemble methods. *Comput. Ind. Eng.* **2018**, *126*, 282–291. [CrossRef]
24. García, M.V.; Irisarri, E.; Pérez, F.; Estévez, E.; Marcos, M. OPC-UA Communications Integration using a CPPS architecture. In Proceedings of the IEEE Ecuador Technical Chapters Meeting (ETCM), Guayaquil, Ecuador, 12–14 October 2016. [CrossRef]
25. Pisching, M.A.; Pessoa, M.A.O.; Junqueira, F.; Filho, D.J.; Miyagi, P.E. An architecture based on RAMI 4.0 to discover equipment to process operations required by products. *Comput. Ind. Eng.* **2018**, *125*, 574–591. [CrossRef]
26. Prada, M.A.; Fuertes, J.J.; Alonso, S.; García, S.; Domínguez, M. Challenges and solutions in remote laboratories. Application to a remote laboratory of an electro-pneumatic classification cell. *Comput. Educ.* **2015**, *85*, 180–190. [CrossRef]
27. Reynard, S.; Gomis-Bellmunt, O.; Sudriá-Andreu, A.; Boix-Aragonés, O.; Benítez-Pina, I. Flexible manufacturing cell SCADA system for educational purposes. *Comput. Appl. Eng. Educ.* **2008**, *16*, 21–30. [CrossRef]
28. Yabanova, I.; Taskin, S.; Ekiz, H. Development of remote monitoring and control system for mechatronics engineering practice: The case of flexible manufacturing system. *Int. J. Elec. Eng. Educ.* **2015**, *52*, 264–275. [CrossRef]
29. Hincapié, M.; Ramírez, M.J.; Valenzuela, A.; Valdez, J.A. Mixing real and virtual components in automated manufacturing systems using PLM tools. *Int. J. Interact. Des. Manuf.* **2014**, *8*, 209–230. [CrossRef]
30. Toivonen, V.; Lanz, M.; Nylund, H.; Nieminen, H. The FMS Training Center—A versatile learning environment for engineering education. *Procedia Manuf.* **2018**, *23*, 135–140. [CrossRef]
31. Pacaux-Lemoine, M.-P.; Trentesaux, D.; Zambrano, G.; Millot, P. Designing intelligent manufacturing systems through Human-Machine Cooperation principles: A human-centered approach. *Comput. Ind. Eng.* **2017**, *111*, 581–595. [CrossRef]
32. Tisch, M.; Hertle, C.; Abele, E.; Metternich, J.; Tenberg, R. Learning factory design: A competency-oriented approach integrating three design levels. *Int. J. Comput. Integr. Manuf.* **2016**, *29*, 1355–1375. [CrossRef]
33. Webpage of the Industrial Engineering School of the University of Extremadura. Available online: eii.unex.es (accessed on 15 September 2018).
34. SMC Webpage. Available online: https://www.smc.eu/portal_ssl/WebContent/main/index_restyling.jsp?lang=en&ctry=EU&is_main=yes&dfl_locale=yes (accessed on 10 September 2018).
35. Ruppert, T.; Jaskó, S.; Holczinger, T.; Abonyi, J. Enabling Technologies for Operator 4.0: A Survey. *Appl. Sci.* **2018**, *8*, 1650. [CrossRef]
36. Zolotová, I.; Papcun, P.; Kajáti, E.; Miskuf, M.; Mocnej, J. Smart and Cognitive Solutions for Operator 4.0: Laboratory H-CPPS Case Studies. *Comput. Ind. Eng.* **2018**, in press. [CrossRef]
37. Esmaeilian, B.; Behdad, S.; Wang, B. The evolution and future of manufacturing: A review. *J. Manuf. Syst.* **2016**, *39*, 79–100. [CrossRef]
38. Zhong, R.Y.; Xu, X.; Klotz, E.; Newman, S.T. Intelligent Manufacturing in the Context of Industry 4.0: A Review. *Engineering* **2017**, *3*, 616–630. [CrossRef]
39. Domínguez, M.; Prada, M.A.; Reguera, P.; Fuertes, J.J.; Alonso, S.; Morán, A. Cybersecurity training in control systems using real equipment. *IFAC-PapersOnLine* **2017**, *50*, 12179–12184. [CrossRef]
40. Calderón, A.J.; González, I.; Calderón, M.; Segura, F.; Andújar, J.M. A New, Scalable and Low Cost Multi-Channel Monitoring System for Polymer Electrolyte Fuel Cells. *Sensors* **2016**, *16*, 349. [CrossRef] [PubMed]
41. González, I.; Calderón, A.J.; Mejías, A.; Andújar, J.M. Novel networked remote laboratory architecture for open connectivity based on PLC-OPC-LabVIEW-EJS integration. Application to remote fuzzy control and sensors data acquisition. *Sensors* **2016**, *16*, 1822. [CrossRef]
42. González, I.; Calderón, A.J.; Andújar, J.M. Novel Remote Monitoring Platform for RES-Hydrogen based Smart Microgrid. *Energy Conv. Manag* **2017**, *148*, 489–505. [CrossRef]
43. Martinez, B.; Vilajosana, X.; Kim, I.H.; Zhou, J.; Tuset-Peiró, P.; Xhafa, A.; Poissonnier, D.; Lu, X. I3Mote: An Open Development Platform for the Intelligent Industrial Internet. *Sensors* **2017**, *17*, 986. [CrossRef] [PubMed]

44. Viegas, C.; Pavani, A.; Lima, N.; Marques, A.; Pozzo, I.; Dobboletta, E.; Atencia, V.; Barreto, D.; Calliari, F.; Fidalgo, A.; et al. Impact of a remote lab on teaching practices and student learning. *Comput. Educ.* **2018**, *126*, 201–216. [CrossRef]

45. Grodotzki, J.; Ortelt, T.R.; Tekkaya, A.E. Remote and Virtual Labs for Engineering Education 4.0. *Procedia Manuf.* **2018**, *26*, 1349–1360. [CrossRef]

46. Mejías, A.; Reyes, M.; Márquez, M.A.; Calderón, A.J.; González, I.; Andújar, J.M. Easy handling of sensors and actuators over TCP/IP Networks by Open Source Hardware/Software. *Sensors* **2017**, *17*, 94. [CrossRef] [PubMed]

47. Bruzzone, A.G.; Longo, F. 3D simulation as training tool in container terminals: The TRAINPORTS simulator. *J. Manuf. Syst.* **2013**, *32*, 85–98. [CrossRef]

48. Scurati, G.W.; Gattullo, M.; Fiorentino, M.; Ferrise, F.; Bordegoni, M.; Uva, A.E. Converting maintenance actions into standard symbols for Augmented Reality applications in Industry 4.0. *Comput. Ind.* **2018**, *98*, 68–79. [CrossRef]

49. González, I.; Calderón, A.J.; Barragán, A.J.; Andújar, J.M. Integration of Sensors, Controllers and Instruments Using a Novel OPC Architecture. *Sensors* **2017**, *17*, 1512. [CrossRef]

50. Fisher, O.; Watson, N.; Porcu, L.; Bacon, D.; Rigley, M.; Gomes, R.L. Cloud manufacturing as a sustainable process manufacturing route. *J. Manuf. Syst.* **2018**, *47*, 53–68. [CrossRef]

© 2018 by the authors. Licensee MDPI, Basel, Switzerland. This article is an open access article distributed under the terms and conditions of the Creative Commons Attribution (CC BY) license (http://creativecommons.org/licenses/by/4.0/).

education
sciences

MDPI

Article

A Systematic Review of Project Allocation Methods in Undergraduate Transnational Engineering Education

Sajjad Hussain [1,*], **Kelum A. A. Gamage** [1], **Md Hasanuzzaman Sagor** [2,*], **Faisal Tariq** [1], **Ling Ma** [2] **and Muhammad Ali Imran** [1]

1   James Watt School of Engineering, University of Glasgow, Glasgow G12 8QQ, UK;
    kelum.gamage@glasgow.ac.uk (K.A.A.G.); Faisal.Tariq@glasgow.ac.uk (F.T.);
    muhammad.imran@glasgow.ac.uk (M.A.I.)
2   School of Electronic Engineering and Computer Science, Queen Mary University of London, London E1 4NS,
    UK; ling.ma@qmul.ac.uk
*   Correspondence: sajjad.hussain@glasgow.ac.uk (S.H.); m.h.sagor@qmul.ac.uk (M.H.S.)

Received: 2 September 2019; Accepted: 17 October 2019; Published: 22 October 2019

check for
updates

Abstract: The final year design project is one of the most important components of any undergraduate engineering program. Fair and efficient project allocation procedures can be vital in ensuring a great student experience and exceptional learning out of these projects, which then could contribute in shaping students' future prospects. In this paper, we review a wide range of project allocation strategies used in various universities at undergraduate levels. We then focus on the project allocations in transnational education (TNE) contexts, which inherit additional allocation challenges. We highlight these challenges and provide recommendations to solve them. We present and compare project allocation strategies adopted at two of the largest TNE programs in China. We also present the factors that influence the project allocations, particularly regarding TNE provisions. Finally, we describe the challenges associated with the project allocations in the TNE scenario, along with proposing some feasible solutions to address these challenges.

Keywords: final year project; undergraduate engineering program; transnational education; project allocations; matching under preferences

1. Introduction

Final year undergraduate projects are generally a key part of any science and engineering degree program. It can be considered the most substantial project undertaken by an undergraduate during their degree, and it largely provides the first opportunity for students to work independently on a project for a year. Throughout the project, students are required to apply the scientific knowledge they gained during their studies, as well as learn new science [1]. Student learning outcomes from final year projects can primarily be divided into two categories: knowledge of science and general skills [2]. For example, knowledge of science refers to the design of an experimental or computational study, as well as problem-solving skills, whereas time management, creative thinking and communicating results can be considered general skills. Final year projects are generally resource-intensive, where the student needs strong academic supervision throughout the project's duration. Achieving a passing grade for the final year project is critical; for example, if a student is unsuccessful to attain a passing grade, they may not even be awarded an accredited degree.

Allocation of projects to students, assessment procedures, access to resources, support and supervision and general management of projects can be considered key influential parameters that govern students learning experiences during the final year project [3]. For example, if students were allocated to projects or supervisors arbitrarily, without considering students preferences or without considering project supervisors' research interests, this will generally result in a poor or unsatisfactory relationship between the student and the supervisor [4]. Project allocation is a resource allocation problem with certain constraints, where to enhance the student learning experience and satisfaction, it is also essential to consider students preferences for projects.

With the prospect of the Teaching Excellence Framework (TEF) on the horizon for universities in the United Kingdom, actions to improve the National Student Survey (NSS) scores have become a vital exercise for many academics. NSS measures student satisfaction from a range of aspects, where it is mainly aimed at final year undergraduate students. The final year project is a vital element of the final year undergraduate programme and it can significantly impact the NSS scores. Consequently, allocating final year undergraduate students to the correct projects is essential for improving their learning experience, engagement and eventually to enhance their satisfaction [5,6].

University academics, particularly in the field of science and engineering, have shown great interest regarding the practices currently used to allocate final year undergraduate students to projects [4,7,8]. The main reason behind that interest is the extreme importance of final year projects and their contribution towards the award of degrees to the students. The project allocations are virtually the starting point of the year-long projects, and therefore it is vital to allocate the projects in a way that is acceptable to both the students and the staff members for a pleasant and enriching experience. As mentioned in Anson and Smith [8], in the case the project does not match student's interest, they may lose the motivation to work on the project, which could result in a low-quality project outcome, and in some cases, poor student–staff relationships. However, the previous works are limited in terms of the generalisation of the findings for specific scenarios like transnational education (TNE) programs.

Allocation of final year undergraduates to projects is a key challenge for TNE programmes. The global growth of TNE is rapidly increasing and becoming an integral part of the internationalisation strategy of most universities [9,10]. Transnational education is generally challenging for science and engineering programmes, particularly to sustain high standards with all the other constraints, and to meet the requirements of accreditation organisations [11]. This becomes substantial when students enter into their final year of study. For example, allocation of final year projects to students should be conducted considering all the constraints associated with TNE. Simultaneously, it is critical to ensure that student learning and engagement during the project is sustained at the highest level and should demonstrate the effectiveness of the selected project allocation mechanism.

With the increasing number of constraints, universities have to rethink the techniques used to allocate final year projects to students while further improving student learning and student satisfaction. This becomes a significant challenge with the increasing number of students in science and engineering TNE programmes. Simultaneously, universities are subjected to increased scrutiny to maintain high standards from accreditation bodies, as well as through the potential TEF. There are numerous final year project allocation techniques published in the literature, where the primary objective of this paper is to review such techniques with the focus of identifying the effectiveness of those techniques with respect to the challenges of TNE. The main objective of this review is to not only provide the available options regarding the allocation of projects, but also present the applicability of some of the working scenarios in TNE programs by comparing and contrasting the pros and cons of the current major practices. At the same time, we present some solutions that are efficient and scalable with the growing size of the TNE programs.

2. Project Allocation Methods

Undergraduate final year projects offer opportunities for students to undertake independent project work and to develop subject-specific and generic skills. It also provides an opportunity for staff to work closely with the student and strengthen individual students' skills, which are not visible from a standard course assessment. However, the success of achieving some of these underlined objectives also depends on the project allocation scheme used at the beginning. Various types of project allocation techniques are used in engineering and science streams: project selection by students based on project titles provided by staff, project allocation based on the preferences of both (or negotiation between) students and lecturers, project based on student's own proposal, etc. Each allocation technique has potential strengths and weaknesses, where a student's choice of project is influenced by their desire to work with a particular academic staff member or desire for a particular project area [7].

Project allocation is a resource allocation problem where the key constraint for any programme is to ensure that the project workload among the staff are distributed evenly while matching projects to student's demands. Doing this for a large number of students is a challenge, where the final year projects are allocated manually.

The manual processing sequence is very time consuming and inconvenient to all parties involved. For instance, a student may have to manually search for a good number of project titles to find the relevant projects and then prioritise them in a form that is difficult to modify after submission. It is also troublesome for supervisors to keep track of the final year project proposals that are submitted and make changes later on. It is also very stressful and tedious for the committee members to manually assign final year projects to students one by one [12,13]. Therefore, most of the universities with a larger number of students these days apply some form of computer algorithms that perform the allocations based on certain inputs and constraints [14,15]. In this section, we will discuss a few of the most common and popular student project allocation (SPA) methods.

2.1. Project Allocation Based on the Preferences of Both (or Negotiation between) Students and Lecturers

This is one of the most common SPA methods, where both students and supervisors have their own preferences. Typically, the available projects are advertised to the students, and having browsed through the descriptions, each student (either explicitly or implicitly) forms a preference list over the projects that they find acceptable. Supervisors may also have preferences over the students and/or the projects that they offer. Manlove and O'Malley studied the problems of allocating projects, where both students and lecturers have preferences over projects, and both projects and lecturers have capacities [16]. They proposed different algorithms and tried to find a more stable allocation process but could not strongly propose one single method without having some approximation. Later, Iwama et al. built upon the algorithms presented by Manlove et al. and proposed an improved stability index for SPA [17]. Moussa and El-Atta also studied the algorithms of Manlove and O'Malley [16] and presented a new SPA model in which the lecturers have preference lists over pairs (student, project), and the students have preference lists over projects [18]. Furthermore, Kazakov [13] mentioned several complexities after analysing two different approaches applied in two consecutive academic sessions, where both the students and supervisors have preferences over projects. Kazakov identified several problems of those methods and proposed a new method having three phases, which saved time for both students and supervisors and reduced the number of randomly allocated projects [13]. Other than making a preferred list of project titles, in some cases, students contact supervisors directly and express their interest in listed projects. A project is allocated to a student if both parties, that is, the student and the supervisor, agreed and confirmed on the same project number. The concept of first come, first served is applied and the process is usually conducted via email. Gallagher et al. suggested that with this system, there is still the problem of "popular" titles, where a large number of students are attracted to a small number of projects [19].

2.2. Project Selection by Students Based on Project Titles

This is another popular SPA system, where students choose their projects by themselves based on the project titles provided by the supervisors. Many higher educational institutions worldwide have adopted this system to allocate final year projects to students [3,7,20,21]. Cheung et al. [12] described the method the Department of Civil Engineering of National University of Singapore follows to allocate the projects to final year students. They proposed an algorithm, which is intended to find an optimal allocation scheme that best matches a student's preference to the student's eligibility for the corresponding project, subject to the constraints in the student's ranking, their prioritised project selections and available project spaces. This allocation scheme ensures that everyone gets a project that best matches the student's personal preference with their ranking.

Harland et al. [7] carried out a case study to compare the factors that influence students' choice of project in the two allocation systems, namely choice of specific title and choice of subject area followed by negotiation, and to determine whether different factors were relevant. This case study demonstrated that there is no significant difference in the factors affecting student project choices between allocation by project title and allocation by subject area followed by individual negotiation. However, the staff were generally much more enthusiastic about allocation by subject area than by title. Open comments indicated that they were able to match students' interests, and to some extent abilities, more closely to the research projects they had available. However, SPA by project titles has benefits, such as it saves the supervisors' time spent on negotiations with students regarding any project. It also ensures that the higher-ranked students get their desired projects to work with.

2.3. Project Selection by Students Based on Supervisors and/or Project Category

Obtaining a satisfying allocation for both students and supervisors by negotiation and/or preference list is a challenging task, especially when the number of available supervisors is small and their popularities are highly diverse. Serrano et al. [22] stated that no allocation system can guarantee that every student gets their first choice when the number of students is significantly greater than the number of available supervisors. They proposed a novel method based on a ranked list of supervisors, as well as categories provided to student, where a category corresponds to a general research area. A student's satisfaction may therefore correspond to getting a project either with a highly ranked supervisor and/or in a highly ranked category. Although they claimed to have an improved level of satisfaction of students and academics, this method could be more time consuming as students have to negotiate the project title with their preferred supervisor or in their preferred area of research even after the allocation. A similar problem will occur in the method proposed by Salami and Mammam [4], where they proposed assigning supervisors to students rather than assigning project titles by using their algorithm. According to them, the advantages of this method is that the projects are not required to be available at the time of allocation, and students and supervisors can discuss their project ideas/topics with each other after the allocation.

2.4. Project Allocation Based on Students' Own Proposals

Another common and popular way of SPA is a "student-led" allocation system. In this system, students design their own project and approach a member of staff to be their supervisor. Students contact supervisors directly via e-mail or in person, and it is up to that member of staff to agree to supervise the student or refer them to someone else. The topic and content of the project is established entirely between the supervisor and student. Thus, a minimum of admin staff support is needed until after the topic and supervisor have been identified [23]. There are some positive aspects of running a student-led model as Harland et al. stated that projects suggested/proposed by students promote active student participation [7]. Chang [24] argues that in this method, independent students need inspiration and occasional guidance rather than full supervision such that students approaching the end of their degrees become autonomous and independent learners. Despite having these advantages, the student-led model also raises a number of issues for the undergraduate cohort as a whole. First, most undergraduates find choosing a research topic difficult as undergraduate students rarely have deep knowledge of any particular area in order to identify a research rationale. Students often identify a very general topic area for research, and usually produce research questions that are too broad to be tenable [23]. In Hidi and Renninger's terms [25], their interest in research needs some substantial external support, and therefore, supervisors need to spend time working on the feasibility of the project, even though students work fairly autonomously on their dissertations. This can end up with the supervisor suggesting a very different topic afterward negotiations, where students can feel disenfranchised as their ideas are set aside and they are channelled into a project for which they have less interest, enthusiasm and ownership [23]. Volkema [26] also suggested that asking students to create their own projects from scratch presents a few difficulties and takes considerable time. Analysis from the literature and informal feedback from those involved suggests that the student-led system is unsatisfactory for the majority of staff and students.

2.5. Other Project Allocation Techniques

Besides the methods described above, many higher education institutes follow other kinds of SPA methods. For instance, algorithms presented in References [20,21,27] provide students with two options, where students can either choose one project from the pool of supervisors' proposed titles or they can propose their own title. Also, many schools/departments, especially outside the U.K., prefer final year projects to be done in groups [3,21]. This can significantly reduce the load for supervisors and make the allocation process a lot easier. However, this method is mainly adopted so that students can learn teamwork and develop communication and leadership skills through their final year project. Allowing group projects can also solve one critical problem of SPA, as described in Section 2.1. As no allocation system can guarantee students first choice when the number of students is significantly higher than the number of projects, it becomes common that more than one student is attracted to the same project. Anwar et al. [28] suggested an alternative strategy, where individual students make a ranked selection of projects as usual, but form groups of up to three members from the individuals, who preferred to get the same project.

3. Project Allocations in TNE Provision through a Project Database

In this section, we will focus on some of the allocation methods deployed by two of the most successful TNE programs in China. One of the programmes is the joint degree program between the University of Glasgow, U.K., and the University of Electronic Science and Technology of China (UESTC), China, namely Glasgow College UESTC (GC-UESTC); and the other program is between Queen Mary University of London (QMUL), U.K., and Beijing University of Posts & Telecommunications (BUPT), China, namely the QMUL-BUPT joint program. The student population in GC-UESTC is 540, while it is 625 in QMUL-BUPT.

3.1. Student–Staff Agreement before Project Allocation: An Offline Method

This approach is followed by the QMUL-BUPT Joint Programme. Each year, all staff members propose 10–11 projects. In June, before the project starts in the new academic year, a final year project workshop to year 3 students is held in Beijing. This is for participating supervisors to introduce themselves and their project ideas to the potential project students. All supervisors' introduction files are also published on QM+, a local portal.

At the beginning of the academic year, supervisors first outline the proposed project ideas, which include a brief description of the project, four tasks, three measurable outcomes, required skills and difficulty level indicator. The project outlines from QMUL supervisors are reviewed by a panel made up of BUPT academic staff and vice versa. Once approved, the project outlines are released to students on QM+. All projects are released at the same time and students can do a search and filter by various options.

The students are given about 3 weeks to study the proposals. Students then contact supervisors to express their interest in a particular project. Supervisors evaluate the competency and appropriateness of the student for the project. This could be done through email exchanges or formal interviews. Once the supervisor and the student mutually agree to work together, the project is removed from the database and a contract is signed between supervisor and student regarding the project allocation. This approach is based on a first come, first served principle. After the allocation deadline, a match between unallocated students and projects is made by the coordinator taking required skills, application area, etc., into account as much as possible.

As can be seen, this allocation method works well since the students have the opportunity of getting their first-choice projects and the staff have the option of selecting students who they consider suitable for the project. Therefore, both students and staff have some sort of control in the project allocations. However, at the same time, the process is time consuming for students and particularly so for staff. For a large student cohort, it may not be feasible for the staff to respond and evaluate all the interested candidates. At the same time, the students have to spend time convincing the staff to assign them the project without the guarantee of a project allocation. Some weak students get neglected in this process as they tend to avoid contacting supervisors due to a lack of confidence in their abilities.

3.2. Project Assignment through a Matching Algorithm: An Online Method

In this section, we discuss the project allocation methodology used by GC-UESTC. Before we proceed further, we provide a brief background about the program model. At GC-UESTC, there are 540 students admitted each year in three degree programs, Electronics and Electrical Engineering, Communication Engineering and Microelectronics System Engineering. After completing all the degree requirements, the students receive two degrees, one from the University of Glasgow and the other from UESTC. Approximately half of the courses are taught by local UESTC staff, while the remaining half of the courses are taught by University of Glasgow staff who fly in from Glasgow to Chengdu for teaching. The block-based teaching model is used where one month's worth of teaching is condensed into one week of teaching. Since Glasgow academics work on a fly in, fly out model, they have limited interaction with the students compared to UESTC staff.

For the project allocations, all staff members from both universities upload their projects on a specifically designed project database. Students are provided 3–4 weeks to study the projects. During this time, they have the option of contacting the supervisors for queries and clarifications. Later, the project database is opened for students to make their selections in the order of their preferences. Students rank their project preferences separately for University of Glasgow staff and UESTC staff. Once all the student preferences are registered, the preference list is used as an input to a matching under preferences algorithm [29,30]. This algorithm is specially designed to match the preferences under a set of conditions. These conditions include the minimum and maximum number of projects a staff member can supervise, the number of students who could work on a project, the target average load of a staff member, the split in load between University of Glasgow and UESTC staff supervision, etc. The objective function of the algorithm could take several forms. Some of the most used objective functions could be maximising the number of students with project allocations matching their highest preferences or minimising the number of students with project allocation matching their lowest preferences.

This algorithm ensures that all the students are assigned the projects according to their top preferences. In some scenarios, the algorithm could fail to converge for some of the students meaning the algorithm is unable to allocate a project to a student from any of the student preferences. The manual selection process can be deployed for the remaining students as the number of unassigned students is small (less than 2%).

This process is extremely efficient in terms of saving time for both staff and students, while ensuring students' preferences are matched to their allocations. The solution also appears transparent and fair to all the staff and student, eliminating student discontentment and staff competitiveness. However, both students/staff are not guaranteed to have the supervisor/student of their own choice. We propose to have a small percentage of pre-allocations in place to address this issue to give some scale of control to staff and students in the project allocations.

3.3. Comparison of the Presented Approaches

In this section, we compare both of methods presented above, which are currently being practiced in two of the largest TNE programs in China. We have presented our findings in the form of a table, Table 1, where we have compared the two allocation processes based on several features.

Table 1. Comparison of the allocation methods in two TNE programs.

Feature \ Institute	QMUL-BUPT	GC-UESTC
Time efficiency	Time consuming due to the staff–student negotiation phase	Time efficient due to the absence of negotiations
Workload efficiency	More workload for staff and students	Little workload on staff and students
Scalability	Less scalable for a large number	Extremely scalable for any number of staff and students
Student control	More student control as there is a possibility of getting their top preferred project	A little less control as the students have a higher possibility of getting one of their lesser preferred projects
Staff control	More staff control due to their role in the selections	No staff control

4. Factors that Affect the Project Allocations in TNE Programs

In this section, we discuss the factors that influence the project allocations. Our main focus is the TNE provision and we reflect on how some factors critically play their roles in the project allocations.

4.1. Student Size

Student size is one of the basic factors that impact the project allocations. With few students to be assigned projects, the allocation problem complexity is generally quite low. However, as the student number grows, the problem complexity grows as well. Meeting the expectations of all the students while taking their interests and supervisors' loads into account, the project allocations become challenging. Rasul et al. [31] stated that the coordination and supervision of final year projects is challenging, especially where large numbers of students work independently with large numbers of supervisors. Limited resources make it difficult for project course coordinators to provide adequate staff development for supervisors. Monitoring the work of so many supervisors is also a difficult task for the project coordinator. Moreover, Johnson and Johnson [32] indicates that the capabilities of students to manage individual project work need to be taught explicitly by the mentor; it is not sufficient just to ask a student to work to manage a project. And according to Rasul et al. [31], supervisors supervising five or more different undergraduate projects at one time have little time to "train" students, which is a very common issue for programmes with a large number of students.

Most of the TNE programs are running with large student populations, especially the TNE programs involving China, which at times exceed 500 in student size. With such large numbers, it may not always be possible to perform the project allocations fairly and efficiently and this leads to the necessity of exploring new approaches in project allocations. In response to increasing student numbers, Healey et al. [33] suggested that group project work could be used to maintain the quality of coordination and supervision. This would also give them experience with teamwork, which is common in many jobs and in some conceptions of research in science and engineering. Another benefit of working in teams is that, where the group functions effectively, better quality work is often obtained than where students work individually. A few case studies are provided in Healey et al. [33] showing how the average mark for the *Issues in Environmental Geography* module based on a group projects was consistently 3–5 percentage points higher than the average for other individual final-year modules in geography.

4.2. Project Proposal Presentation

It is important to consider how the project proposals are presented to the students. In some scenarios, especially during the 1990s, the staff used to upload their projects through some student portals or on their personal websites [13], while in some cases, the list of offered projects used to be displayed at their office doors or notice boards. Students could discuss projects with supervisors and then sign a form to confirm a written agreement. This is also the case at some universities these days where the total student number is not that high. In some institutions, the allocation of students to the supervisor is done by the head of the academic department or the project coordinator. Students then meet their supervisors and the project topics are given to the students to work on or the students themselves submit a list of topics to the supervisors and the supervisors choose their preferred one [34]. However, this approach is characterised by two or more problems, namely biasedness and conflict of interest, as noted by Aderanti et al. [34]. To accommodate a large student population, universities have developed their own project databases where all the staff upload their proposals to the project allocation interface and all proposals are available for the student to view and make their selections. As discussed earlier, several models have been proposed by education researchers for this selection process, which can satisfy both students and supervisors, as well as the project coordinator. For example, Abraham et al. [29] suggested a model where the students supply preference lists over projects that were offered by lecturers and each lecturer supplies a preference list over students who show interest in one or more of their projects. Li Pan et al. developed a model using goal programming [35], where they tried to maximise the number of assigned projects, satisfying as much as possible both students' and the department's preferences. Dye [36] and Kazakov [13] proposed models using a "stable marriage" algorithm to match members from two different lists (e.g., men and women, students and projects, etc.) according to the preferences expressed by each of the lists' members. Harper et al. [37] and Srinivasan and Rachmawati [38] also proposed evolutionary algorithms to solve the project allocation problem. Recently, Chiarandini et al. has studied the problem of allocating students into teams working on project topics and proposed a new allocation model using a state-of-the-art commercial solver [39].

Besides all those proposed allocation methods, there are a few things we would like to suggest that could be done to facilitate students in their project selection process. For instance, dividing the projects into main research themes would allow the students to browse only the projects of their research interest and thus would not only save their time but also help them focus more on the topics of their interest instead of random project browsing. Moreover, even inside a specific theme, the sequence in which the projects appear to the students plays an important role in their selection-making process. If a student has to go through 50–100 projects, the level of their interest would drop significantly for the projects that appear later on in the list. Therefore, it is recommended to make the project display sequence random such that each student sees the projects in a different order. This will ensure that all projects are given equal attention by the students. This will also help in balancing the load among different staff members. Finally, the project titles and descriptions should be as concise as possible to let the students browse more projects in less time while keeping their focus intact.

4.3. Project Themes

The project allocations are very much related to the project themes offered to the students. Some themes might be of more interest to the student than others. This could disturb the balance between different themes, and subsequently, the staff responsible for supervising the projects related to those themes. Therefore, it is recommended that the main project themes should be selected carefully where one clever way of doing so could be the alignment of the projects with the modules that students have previously taken in their degree program. It is probable that students tend to work on the projects and feel confident in selecting the topics for which they have some previous background knowledge and do not feel comfortable with topics unfamiliar to them. A common comment of examiners about final year projects is that students do not appear to have been adequately prepared. This is confirmed by research into the student experience of dissertations by Todd et al. [40]. Another way of making students well prepared is to start preparing students for their final year project from the day they start their degree in higher education. Healey and Jenkins [41] described the pedagogical shift that is beginning to occur through examples of degree programmes, where students are introduced to research from the start of their degree. It is worth mentioning that over 80% of students at the Massachusetts Institute of Technology (MIT) undertake at least one undergraduate research opportunity programme, often in their junior year, mostly in addition to their studies, according to the report by Huggins et al. [42]. Therefore, effort should be made to synchronise the project themes with the major learning outcomes of the previous courses taken, as well as start preparing students for their final year project from the very beginning of their degree programme. This would result in better student performance on their projects, as well as a lower supervision load for the staff.

4.4. Supervisor's Profile

It is one of the important factors when it comes to the project allocations that some of the staff members are relatively more popular among students than others. There could be several reasons for that including the supervisor's research profile, the research group, good teaching skills, friendly student rapport, etc. According to the students' responses presented in Stefani et al. [43], the primary expectation of a student from their supervisor is "to ensure the student is on the right track, offering constructive criticism and offering guidance when necessary". It is also evident from the report of Todd et al. [40] that supervisors who provide constructive feedback on draft work are highly appreciated. The report also argued that approaches adopted by tutors towards supervision varied greatly in terms of formality. Certain supervisors had a relatively relaxed approach, initiating preliminary meetings but then leaving it to the student to request support when needed, with others being more formal and directive, e.g., establishing a supervision timetable for the year; producing a written record of each meeting and drawing up a formal contract of rights and responsibilities. Although the majority of interviewed students thought that the latter approach was useful for them, there were many who preferred supervisors adopting the former approach.

In academic environments, it is observed that such student preferences create healthy competition among staff members; however, the development of enviousness for popular staff members is also produced. It should also be noted that there is a student preference divide for local supervisors versus foreign supervisors in the context of TNE programs. This impacts the student preferences and therefore staff loading. We recommend avoiding such issues through incorporating appropriate constraints in the allocation strategies.

4.5. Industrial Projects/Community-Based Learning

With students having ambitions to join top industries after graduation, the industrial projects could pave their way to the doors of those industries. While the academia–industrial collaboration is important in many aspects, final year projects could play a critical role in strengthening that relationship. There is considerable evidence that student engagement with external employers benefits their learning while making a practical contribution to communities and companies. Mason O'Connor et al., for example, note that the literature on community engagement through the curriculum suggests it enhances the quality of academic work, employability and lifelong learning [44]. Moreover, Lee et al. claim that external engagement with real issues has been shown to increase students' confidence through placing them in positions of responsibility and exposing them to a greater diversity of learning experiences [45]. There are other benefits of engaging in community and work-based learning for students, such as the development of critical thinking, gaining insight into the complex nature of knowledge, showing enthusiasm for a subject and greater subject-related understanding [46,47]. For TNE programs involving China, there are huge industrial project opportunities that should be explored and the students should be provided with the option to work on industrial projects.

4.6. Staff Load Balancing

The project allocations are hugely impacted by the supervision load a staff member could be assigned. With a small student number, this issue may seem irrelevant, but with hundreds of project supervisions to be done each year in TNE programs, the staff supervision load balancing becomes critical. According to MacKeogh [48], the traditional role of the supervisor is to provide guidance, advice, instruction, encouragement and support; however, it is also the supervisor's role to assist students in their management of conflicts and risk. On top of that, the supervisor acts as the student's examiner, providing formative and summative feedback throughout the learning process. This traditional role can place huge pressures on academics and overload them when they are supporting a large number of students. Hensel and Paul express their concern noting that undergraduate research often takes place outside the curriculum; therefore, recognition of the time that supervision takes may be an issue [49]. As such, it is vital to ensure that no staff member is overloaded while the student project preferences are also met with reasonable success. Moreover, it is required that the supervision load balance between the local staff and foreign staff is determined for the smooth implementation of allocation procedures in TNE provision.

5. Challenges and Solutions in TNE Project Allocations

In all the TNE programs, the final year project allocations process comes with additional challenges and constraints. Working transnationally is likely to raise challenges that stem from differences in culture, educational background and expectations. Transnational teachers find themselves living and working (albeit temporarily) within environments that are culturally different to their own. Their initial interactions are likely to engender "culture shock" [50]. They might find that some of these cultural differences have an impact on how they teach within the overseas setting.

Teacher burnout is a serious risk within transnational education. Many TNE teachers report extreme tiredness fuelled by lengthy international travel [51], jet lag and intensive teaching patterns. TNE faculty members are often pushed to their physical limits and still have to enter the classroom and perform professionally. Smith [50] explains this as *"you're literally flying in, your eyes are shutting and then you're having to teach"*.

One of the most challenging aspects of delivery of flying-faculty teachers is that contact is often based on very short and intensive teaching blocks [52]. As such, teachers may have to work through new models of teaching that enable the material to be covered in a much shorter period of time. It is important to mention here that this kind of intensive block teaching is often criticised for having little to do with pedagogy and more to do with convenience [53]. After delivering a week of intensive sessions, many TNE teachers fly back to the home country and start teaching there from the next day. Debowski [54] reports that one of the most difficult things about working transnationally is managing workloads in two locations.

Language can be another challenge for academics working transnationally. Many students often struggle to keep up with the reading that is required, or have difficulties expressing themselves orally [55]. Therefore, teachers need to be aware of their own language: speaking clearly and not too quickly, explaining key concepts using simple words and leaving time at the end of sessions for students to ask questions.

While student engagement is a critical component of any TNE program, this engagement becomes even more important for final year project supervisions. The project allocations should consider this point and make sure that student engagement is not compromised due to the allocation outcomes. Since for most TNE programs, like GC-UESTC and QMUL-BUPT, the non-local staff operate using a fly in, fly out model and perform block-based teaching, there may be fewer opportunities for activities and face-to-face contact with project students. These challenges could be resolved through the use of technology where, for example, the staff are recommended to integrate the contemporary communication tools like Skype, WhatsApp, WeChat (a popular connectivity tool in China), etc., in their day-to-day communication with students to provide a healthy supervision experience to the students. However, Augustsson and Jaldermark note that online supervision requires a different skill set, as online supervision relies mainly on written communication around electronic drafts [56].

At the same time, the role of a second supervisor also becomes important. It is recommended that for all the non-local first supervisors, a local second supervisor is allocated. This allocation must be

Educ. Sci. **2019**, *9*, 258

done carefully in order to match the research interests of both first and second supervisors. Thus, the local staff could contribute efficiently to student supervision in the absence of the non-local staff.

It is a possibility that students may have a tendency to prefer the staff from one partner university while making their selections due to several reasons. For instance, the global visibility of the staff, on-campus availability of the staff, lack of communication skills in a non-native language, etc. While student preferences could become biased towards one side or the other, the allocation procedures should be defined in a way to restrict such biases. One of the possible solutions to this problem could be making sure that the students include projects from both sides in their preference list and then assigning appropriate weight to the student preferences. This would take the student preferences into account as well as keep the balance in the supervision numbers for the TNE partner institutions.

6. Conclusions

In this paper, an overview of final year project allocation strategies and procedures is presented. The focus of this paper is on the context and related challenges of final year project allocations in the TNE provision with large student number supervision requirements. We have presented the details of project allocation procedures for two extremely successful TNE programs. While student staff pre-agreement before allocations provides control to staff and students regarding project allocations, it may not be a scalable solution for a large student population, which is a common feature of many TNE programs. On the other hand, the matching algorithm under student preferences happens to be a scalable, fair and efficient allocation process. However, in order to accommodate staff and student involvement, a pre-allocation process at a relatively small scale is recommended prior to the matching algorithm's execution. We have also presented the factors that are important to consider while allocating projects for TNE programs. Towards the end, we have highlighted some of the challenges associated with the TNE programs and the project allocations in the TNE context. We have also made recommendations to improve the final year project experience for both staff and students generally, and to improve the project allocation procedures specifically.

Author Contributions: Conceptualization, S.H. and M.A.I.; methodology, F.T. and K.A.A.G.; software, K.A.A.G.; validation, M.A.I., S.H.; formal analysis, M.H.S.; investigation, S.H.; resources, L.M.; data curation, S.H.; writing—original draft preparation, K.A.A.G. and M.H.S.; writing—review and editing, F.T. and M.H.S; visualization, L.M.; supervision, S.H.; project administration, M.A.I.; funding acquisition, S.H., K.A.A.G., L.M.

Funding: This research received no external funding.

Conflicts of Interest: The authors declare no conflicts of interest.

References

1. Hunter, A.; Laursen, S.; Seymour, E. Becoming a scientist: The role of undergraduate research in students' cognitive, personal, and professional development. *Sci. Educ.* **2007**, *91*, 36–74. [CrossRef]
2. Ryder, J. What can students learn from final year research projects? *Biosci. Educ.* **2004**, *4*, 1–8. [CrossRef]
3. Teo, C.Y.; Ho, D.J. A Systematic Approach to the Implementation of Final Year Project in an Electrical Engineering Undergraduate Course. *IEEE Trans. Educ.* **1998**, *41*, 25–30. [CrossRef]
4. Salami, H.O.; Mamman, E.Y. A Genetic Algorithm for Allocating Project Supervisors to Students. *Int. J. Intell. Syst. Appl.* **2016**, *10*, 51–59. [CrossRef]
5. Kuh, G.D. The national survey of student engagement: Conceptual and empirical foundations. *New Direct. Inst. Res.* **2009**, *141*, 5–20. [CrossRef]
6. Barber, T.; Timchenko, V. Student-specific projects for greater engagement in a computational fluid dynamics course. *Australas. J. Eng. Educ.* **2011**, *17*, 129–138. [CrossRef]
7. Harland, J.; Pitt, S. Venetia Saunders Factors affecting student choice of the undergraduate research project: Staff and student perceptions. *Biosci. Educ.* **2005**, *5*, 1–19. [CrossRef]

8. Anson, R.A.I.; Smith, K.A. Undergraduate Research Projects and Dissertations: Issues of topic selection, access and data collection amongst tourism management students. *J. Hosp. Leis. Sport Tour. Educ.* **2004**, *3*, 19–32. [CrossRef]

9. Naidoo, V. Transnational higher education: A stock take of current activity. *J. Stud. Int. Educ.* **2009**, *13*, 310–330. [CrossRef]

10. Naidoo, V. Transnational higher education: Why it happens and who benefits? *Int. High. Educ.* **2010**, *58*. [CrossRef]

11. Miliszewska, I.; Horwood, J.; McGill, A. Transnational Education through Engagement: Students Perspective. In Proceedings of the Informing Science and IT Education Conference, Pori, Finland, 24–27 June 2003; pp. 165–173.

12. Cheung, Y.; Hong, G.M.; Ang, K.K. A dynamic project allocation algorithm for a distributed expert system. *Expert Syst. Appl.* **2004**, *26*, 225232. [CrossRef]

13. Kazakov, D. Coordination of Student Project Allocation, University of York. Available online: http://www-users.cs.york.ac.uk/kazakov/papers/proj.pdf (accessed on 13 February 2018).

14. Ismail, S.I.; Abdullah, R.; Kar, S.A.C.; Fadzal, N.; Husni, H.; Omar, H.M. Online project evaluation and supervision system (oPENs) for final year project proposal development process. In Proceedings of the 2017 IEEE 15th Student Conference on Research and Development, Putrajaya, Malaysia, 13–14 December 2017; pp. 210–214.

15. Abdulkareem, A.; Adewale, A.; Dike, I. Design and development of a University portal for the management of final year undergraduate projects. *Int. J. Eng. Comput. Sci.* **2013**, *2*, 2911–2920.

16. Manlove, D.; O'Malley, G. Student-project allocation with preferences over projects. *J. Discret. Algorithms* **2008**, *6*, 553–560. [CrossRef]

17. Iwama, K.; Miyazaki, S.; Yanagisawa, H. Improved approximation bounds for the Student-Project Allocation problem with preferences over projects. *J. Discret. Algorithms* **2012**, *13*, 59–66. [CrossRef]

18. Moussa, M.I.; El-Atta, A.H.A. A Visual Implementation of Student Project Allocation. *Int. J. Comput. Theory Eng.* **2011**, *3*, 178–184. [CrossRef]

19. Gallagher, A.M.; George, M.; Gill Chris, I. *Streamlining Allocation and Assessment of Traditional Final Year Research Projects across Multiple Undergraduate Degree Programmes*; Centre for Bioscience, The HE Academy: York, UK, 2008.

20. Hasan, M.H.; Sahari, K.S.M.; Anuar, A. Implementation of a New Preference Based Final Year Project Title Selection System for Undergraduate Engineering Students in UNITEN. In Proceedings of the 2009 International Conference on Engineering Education (ICEED 2009), Kuala Lumpur, Malaysia, 7–8 December 2009; pp. 230–235.

21. Pudaruth, S.; Bhugowandeen, M.; Beepur, V. Multi-Objective Approach for the Project Allocation Problem. *Int. J. Comput. Appl.* **2013**, *69*, 26–30. [CrossRef]

22. Calvo-Serrano, R.; Guillén-Gosálbez, G.; Kohn, S.; Masters, A. Mathematical programming approach for optimally allocating students' projects to academics in large cohorts. *Educ. Chem. Eng.* **2017**, *20*, 11–21. [CrossRef]

23. Knight, R.-A.; Botting, N. Organising undergraduate research projects: Student-led and academic-led models. *J. Appl. Res. High. Educ.* **2016**, *8*, 455–468. [CrossRef]

24. Chang, H. Turning an undergraduate class into a professional research community. *Teach. High. Educ.* **2005**, *10*, 387–394. [CrossRef]

25. Hidi, S.; Renninger, K.A. The four-phase model of interest development. *Educ. Psychol.* **2006**, *41*, 111–127. [CrossRef]

26. Volkema, R.J. Designing effective projects: Decision options for maximising learning and project success. *J. Manag. Educ.* **2010**, *34*, 527–550. [CrossRef]

27. Lei, W. A Web-Based Project Allocation System. Available online: https://pdfs.semanticscholar.org/c827/d370e0affa4b76b5a3a0a789dcb12a5aefa8.pdf (accessed on 15 February 2019).

28. Anwar, A.A.; Bahaj, A.S. Student project allocation using integer programming. *IEEE Trans. Educ.* **2003**, *46*, 359–367. [CrossRef]

29. Abraham, D.J.; Irving, R.W.; Manlove, D.F. Two algorithms for the Student-Project Allocation problem. *J. Discret. Algorithms* **2007**, *5*, 73–90. [CrossRef]

30. Manlove, D. *Algorithmics of Matching under Preferences*; World Scientific: Singapore, 2013.

31. Rasul, M.; Nouwens, F.; Martin, F.; Greensill, C.; Kestell, D.S.C.; Hadgraft, R. Good practice guidelines for managing, supervising and assessing final year engineering projects. In Proceedings of the 20th Australasian Association for Engineering Education Conference: Engineering the Curriculum, Adelaide, Australia, 6–9 December 2009; pp. 205–210.
32. Johnson, D.; Johnson, R. *Learning Together and Learning Alone: Cooperation, Competition and Individualization*, 5th ed.; Allyn & Bacon: Boston, MA, USA, 1984.
33. Healey, M.; Laura, L.; Arran, S.; James, D. *Developing and Enhancing Undergraduate Final-Year Projects and Dissertations*; Higher Education Academy: York, PA, USA, 2013.
34. Aderanti, F.A.; Amosa, R.T.; Oluwatobiloba, A.A. Development of Student Project Allocation System Using Matching Algorithm. *Int. Conf. Sci. Eng. Environ. Technol.* **2016**, *1*, 153–160.
35. Pan, L.; Chu, S.; Han, G.; Huang, J. Multi-criteria student project allocation: A case study of goal programming formulation with dss implementation. In Proceedings of the 8th International Symposium on Operations Research and Its Applications (ISORA'09), Zhangjiajie, China, 20–22 September 2009; pp. 75–82.
36. Dye, J. *A Constraint Logic Programming Approach to the Stable Marriage Problem and Its Application to Student-Project Allocation*; BSc Honours Project Report; University of York, Department of Computer Science: York, UK, 2001.
37. Harper, P.R.; de Senna, V.; Vieira, I.T.; Shahani, A.K. A genetic algorithm for the project assignment problem. *Comput. Oper. Res.* **2005**, *32*, 1255–1265. [CrossRef]
38. Srinivasan, D.; Rachmawati, L. Efficient Fuzzy Evolutionary Algorithm-Based Approach for Solving the Student Project Allocation Problem. *IEEE Trans. Educ.* **2008**, *51*, 439–447. [CrossRef]
39. Chiarandini, M.; Fagerberg, R.; Gualandi, S. Handling preferences in student-project allocation. *Ann. Oper. Res.* **2019**, *275*, 39–78. [CrossRef]
40. Todd, M.; Bannister, P.; Clegg, S. Independent inquiry and the undergraduate dissertation: Perceptions and experiences of final-year social science students. *Assess. Eval. High. Educ.* **2004**, *29*, 335–355. [CrossRef]
41. Healey, M.; Jenkins, A. *Developing Undergraduate Research and Inquiry*; Higher Education Academy: York, UK, 2009.
42. Huggins, R.; Jenkins, A.; Scurry, D. *Undergraduate Research in Selected US Universities: Report on US Visit–Institutional Case Studies*; Higher Education Academy: York, UK, 2007.
43. Stefani, L.A.J.; Tariq, V.N.; Heylings, D.J.A.; Butcher, A.C. A Comparison of Tutor and Student Conceptions of Undergraduate Research Project Work. *Assess. Eval. High. Educ.* **1997**, *22*, 271–288. [CrossRef]
44. O'Connor, K.M.; McEwen, L.; Owen, R.; Lynch, K.; Hill, S. *Literature Review: Embedding Community Engagement in the Curriculum: An Example of University–Public Engagement*; University of Gloucestershire, National Co-ordinating Center for Public Engagement: Bristol, UK, 2011.
45. Lee, G.; McGuiggan, R.; Holland, B. Balancing student learning and commercial outcomes in the workplace. *High. Educ. Res. Dev.* **2010**, *29*, 561–574. [CrossRef]
46. Burack, C.; Prentice, M.; Robinson, G. Assessing service learning outcomes for students and partners. In: International Association for Research on Service-Learning and Community Engagement. In Proceedings of the 10th Annual Conference, Indianapolis, IN, USA, 28–30 October 2010.
47. Deeley, S.J. Service-learning: Thinking outside the box. *Act. Learn. High. Educ.* **2010**, *11*, 43–53. [CrossRef]
48. MacKeogh, K. Supervising undergraduate research using online and peer supervision. In Proceedings of the 7th International Virtual University Conference, Bratislava, Slovakia, 14–15 December 2006; Huba, M., Ed.;
49. Hensel, N.; Paul, E.L. *Faculty Support and Undergraduate Research: Innovations in Faculty Role Definition, Workload and Reward*; Council on Undergraduate Research: Washington, DC, USA, 2012.
50. Smith, K. Exploring flying faculty teaching experiences: Motivations, challenges and opportunities. *Stud. High. Educ.* **2014**, *39*, 117–134. [CrossRef]
51. Gribble, K.; Ziguras, C. Learning to teach offshore: Pre-departure training for lecturers in transnational programs. *High. Educ. Res. Dev.* **2003**, *22*, 205–216. [CrossRef]
52. Leask, B. Teaching for learning in the transnational classroom. In *Teaching in Transnational Education: Enhancing Learning for Offshore International Students*; Dunn, L., Wallace, M., Eds.; Routledge: London, UK, 2008; pp. 120–132.
53. Davies, M.W. Intensive Teaching Formats: A review. *Issues Educ. Res.* **2006**, *16*, 1–20. Available online: http://www.iier.org.au/iier16/davies.html (accessed on 24 September 2019).

54. Debowski, S. Across the divide: Teaching a transnational MBA in a second language. *High. Educ. Res. Dev.* **2005**, *24*, 265–280. [CrossRef]
55. Bodycott, P.; Walker, A. Teaching abroad: Lessons learned about inter-cultural understanding for teachers in higher education. *Teach. High. Educ.* **2000**, *5*, 79–94. [CrossRef]
56. Augustsson, G.; Jaldermark, J. Online supervision: A theory of supervisors' strategic communicative influence on student dissertations. *High. Educ.* **2014**, *67*, 19–33. [CrossRef]

© 2019 by the authors. Licensee MDPI, Basel, Switzerland. This article is an open access article distributed under the terms and conditions of the Creative Commons Attribution (CC BY) license (http://creativecommons.org/licenses/by/4.0/).

education sciences

MDPI

Article

Student's Perceptions Regarding Assessment Changes in a Fluid Mechanics Course

Teresa Sena-Esteves [1], Cristina Morais [1], Anabela Guedes [1], Isabel Brás Pereira [1],
Margarida Marques Ribeiro [1], Filomena Soares [2] and Celina Pinto Leão [3,*]

1   Department of Chemical Engineering, CIETI Research Centre, Polytechnic of Porto, 4200-072 Porto, Portugal;
     mte@isep.ipp.pt (T.S.-E.); lcm@isep.ipp.pt (C.M.); afg@isep.ipp.pt (A.G.); imp@isep.ipp.pt (I.B.P.);
     mgr@isep.ipp.pt (M.M.R.)
2   Department of Industrial Electronics, Algoritmi Research Centre, University of Minho, Campus de Azurem,
     4800-058 Guimarães, Portugal; fsoares@dei.uminho.pt
3   Department of Production Systems, Algoritmi Research Centre, University of Minho, Campus de Azurem,
     4800-058 Guimarães, Portugal
*   Correspondence: cpl@dps.uminho.pt; Tel.: +351-253-510-180

check for
updates

Received: 10 May 2019; Accepted: 12 June 2019; Published: 19 June 2019

Abstract: The main objective of this study is to evaluate students' perceptions regarding different methods of assessment and which teaching/learning methodologies may be the most effective in a Fluid Transport System course. The impact of the changes in the assessment methodology in the final students' grades and attendance at theoretical classes is also analysed, and the results show that students' attendance at theoretical classes changed significantly. The students prefer and consider more beneficial for their learning assessment through several questions/problems and small tests during theoretical lessons instead of a single moment of evaluation. For them, the traditional teaching/learning methodology is still considered the most effective one. At the same time, students perceive that the development of the Practical Work (PW) and several moments of assessment had positive repercussions on the way they focus on the course content and keep up with the subjects taught, providing knowledge on the area under study, encouraging collaborative work and stimulating the students' intellectual curiosity. Largely, students agree that the PW is an important tool in their learning process and recommend it as a teaching activity. In general, students are confident with the knowledge acquired with the PW and feel able to size fluid transport systems.

Keywords: Fluid Mechanics; Students Perceptions; Engineering Education; Assessment; Practical Work

1. Introduction

Several studies can be found in the literature focused on designing and testing new (or reformulated) teaching/learning and assessment methodologies applied in different areas. Most of them have as a final goal promoting class attendance and improving students' performance and knowledge.

Cansino et al. [1] attempted to find out if the attendance at classes and the promotion of students' proactivity through oral presentations of case studies positively affected the probability of passing the exam and obtaining higher scores. In fact, they concluded that class attendance and case study presentations were related with academic performance. Also, Lukkarinen et al. [2] investigated the relationship between voluntary class attendance and learning performance. The authors divided students into three distinct groups: students who attended classes but dropped out before the final exam, students who attended classes and the exam, and students who did not attend classes, studied independently, and attended the exam. Although students may succeed without attending classes, the authors concluded that there was a key group of students for whom class attendance was positively related to performance.

Ma et al. [3] explored the use of the application-based flipped classroom (APP-FC), an innovative teaching-learning model, in an immunology course medical curriculum at Lanzhou University, China. The authors concluded that most students that implemented the APP-FC model improved their score in the final examination (60%) compared to the control group (40%). The majority of students (70%) preferred this teaching model to the traditional lecture-based method. Also, APP-FC improved students' learning motivation, self-directed learning skills, and problem-solving abilities.

There is a growing interest in practice-based learning in countries with both advanced and developing economies. Much of this interest is directed towards augmenting students' learning within vocational or higher education programmes of initial occupational preparation or those for professional development (i.e., further development of occupational knowledge across working life) [4].

Fluid Mechanics (FM) in engineering courses is not an exception. Several works can be found in the literature [5–10]. In fact, FM is considered a challenging topic, leading some of the times to unsuccessful students and distressed teachers.

Since 2013, in the Curricular Unit Fluid Mechanics at the University of Western Sydney (Australia) a Blended Learning Approach (BLA) has been employed. This teaching/learning approach considers online recorded lectures and tutorials, hand-written tutorial solutions, discussion board and online practice quizzes. The tutorial classes are devoted to discussing difficult topics and engaging students with practical applications. Students' feedback is positive, as the learning experience improved, and the knowledge of fluid mechanics translated into a higher average grade and an increase in completion rates [9].

Nevertheless, the implementation of new teaching/learning methods can be a hard task for students as well as for teachers; it is well known that any change in the traditional and commonly accepted teaching/learning process may cause certain resistance from both. If correctly implemented, problem/project-based learning methodologies, case studies, and hands-on approaches (among other interactive processes, where students are the main actors) are key tools to prepare students for their engineering profession [11–15]. These tools allow students to develop their ingenious, social and management skills, which are undoubtedly relevant for engineers in their profession [16].

1.1. Motivation

Apart from the implementation of new/different teaching/learning tools, students' assessment is also a topic to explore in higher education. Gibbs and Simpson [17] refer to two important aspects: giving feedback to students on their learning and adequate students' workload for the assessment method.

Sorensen [8] presents a study on second year students' perceptions regarding the assessment in a traditional module conducted in the Department of Chemical Engineering at University College London. In this particular case, the assessment was performed through online quizzes that give immediate feedback to students, allowing them to identify the contents that might need to be improved. The analysis of questionnaires made it possible to infer that students react very positively to this assessment mode, suggesting that this method be extended to other modules.

An interactive online system for skills and knowledge assessment in a computer engineering course and its impact on the students' learning process was presented by Hettiarachchi et al. [18]. Through data analysis, the authors concluded that this type of assessment had a positive impact on students' learning and performance. In fact, students were engaged in the interactive system for both practices, where they were able to evaluate their progress or difficulties, and also in the assessment process.

The authors believe in the importance of promoting student self-learning and adapting the assessment methods in order to prepare them for their professional challenges. Following this trend, the study presented in this paper is focused on how the assessment in a Fluid Mechanics course is perceived by students. How different evaluation methods are considered relevant by students? How can the evaluation methods help students in their future profession?

1.2. Objectives

The main objective of this study is to evaluate students' perceptions regarding different types of assessment. In addition, the impact of the changes in the assessment methodology in the final students' grades is analysed. These analyses will allow the identification and adoption of the teaching/learning methodologies that may be the most effective.

To fulfil the main objective, a main research question (RQ) was defined:

RQ—How do students perceive the assessment methodology?

Moreover, in order to analyse the effect of introducing a Practical Work (PW), five sub-research questions were formulated:

sRQ1—How important is the PW in the students' learning process?

sRQ2—Is the PW perceived as an autonomous tool?

sRQ3—Is the PW effective for preparing students for the challenge of professional life?

sRQ4—Which competences did students acquire with the PW?

sRQ5—Are the students confident with the knowledge acquired with the PW to size a proper fluid transport system?

It is the authors' belief that the optimisation of the assessment methodology may have a positive impact in preparing the students for the challenge of their professional learning.

The paper is organised in four sections. In the next section (Materials and Methods), the curricular unit (CU) Fluid Transport Systems (with the acronym STFLU) is characterised followed by the assessment methodology definition, by the questionnaire used to evaluate students' satisfaction and perceptions and by the students' characterisation. In the third section, the analysis of students' perceptions is made, and in the last section, the final remarks are addressed.

2. Materials and Methods

In this section, the curricular unit Fluid Transport Systems (STFLU) is characterised, followed by the assessment methodology definition. The questionnaire used to evaluate students' satisfaction and perceptions is also described, followed by the students' characterisation.

2.1. Fluid Transport Systems in Chemical Engineering

Since 2006, the Chemical Engineering graduation in the Higher Education Institution (HEI) has been divided into two cycles, the First Cycle (3 years) and the Master degree (2 years). They represent two complementary levels of higher education. The first cycle enables students to gather technical and scientific tools that will allow them to act in industry, quality control and laboratory services. The Master degree provides a deepening of knowledge in the fundamental areas of chemical engineering and offers specific subjects in each of the three branches presently available: environmental protection technologies, energy and biorefinery, and quality. The present study refers to the Fluid Transport Systems (STFLU) course, which is taught in the 1st cycle.

2.1.1. Course Characterisation

STFLU is a course of the second year of the first cycle in Chemical Engineering of HEI. Since 2012/13, this course has had a total of four hours per week (one hour of lectures and three hours of practical classes) in daytime and evening classes. On average, there are 55 to 65 students each year, and two teachers. Theoretical (T) classes are mainly expositive but the teachers also use the interrogative method and different demonstrative and active techniques. In theoretical-practical (TP) classes, students are requested to cooperatively solve problems and real-world case studies.

2.1.2. Course Objectives

STFLU has the general objective of giving students fundamental knowledge in Fluid Mechanics that will enable them to design fluid transport systems and select the associated equipment. In particular, at the end of this course, students should be able to make mass and energy balances that are necessary to design systems and select the appropriate equipment (flow meters; centrifugal, reciprocating and rotary pumps; compressors and fans).

2.1.3. Course Syllabus

The syllabus of STFLU course is divided into two main parts. The first one covers the fundamental principles of mass, energy and momentum transport and the second parts includes the systems of fluid transport (selection and sizing of pipes, valves and fittings, flow meters, pumps, compressors and fans).

2.1.4. Assessment Methodology

The assessment methodology of this course has been changed in recent years. This change should improve the knowledge acquisition by the students and consequently it should be reflected in the final students' grades and success rate.

Figure 1 shows the last assessment flowchart, corresponding to the academic year 2017/2018, which reflects the evaluation towards an autonomous/independent learning by students.

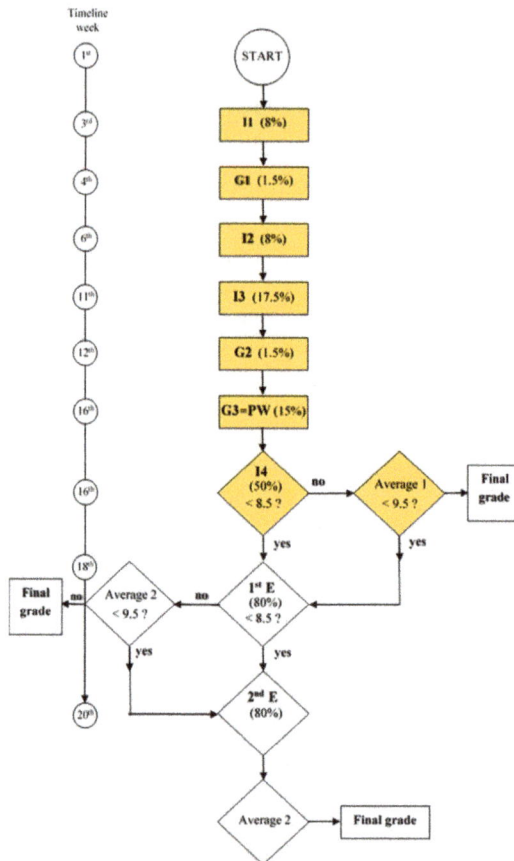

Figure 1. Assessment flowchart for the academic year 2017/18 (I stands for Individual and G for Group evaluation moments).

The assessment methodology for the academic years between 2006/07 and 2015/16 included only PW (at the Moment G3) and the two final exams (normal and supplementary periods of exams), with the exception of 2009/10, when students only had the two final exams (without the PW).

For the last two academic years (2016/17 and 2017/18), students could choose between two modes of assessment:

- four individual mini-tests (at Moments I1, I2, I3 and I4), one of the two group tests (Moments G1 and G2, from which they can choose the one with the best grade), and the PW (Moment G3) or
- a Final Exam (FE) and the PW (Moment G3).

In both cases, the PW had a weight of 15% in the final grade. In case students decide to take the FE, this had a weight of 85%. Otherwise Moment I1 had a weight of 8%, I2 had a weight of 8%, I3 had a weight of 17.5%, I4 had a weight of 50% and Moment G1 or G2 had a weight of 1.5% on the final grade.

There was no minimum score in 2016/17 in any of the assessment components but in 2017/18 a minimum score of 8.5 out of 20 was demanded in moment I4 and in the final exam (normal period of exams).

2.2. Students' Perceptions Questionnaire about Fluid Transport Systems

To fulfil the main objective of this study, the impact of the changes regarding knowledge assessment changes in a Fluid Mechanics course, it is important to analyse the students' opinion. Therefore, a questionnaire was developed [19,20] to obtain the students' perceptions not only regarding the assessment methodology, but also identifying teaching/learning methodologies.

2.2.1. Design of the Questionnaire

The questionnaire was based on a previous one that studied the impact of introducing a PW in the learning process of the Fluid Transport Systems course in the Chemical Engineering degree [20]. Briefly, the questionnaire comprises the following parts:

1. Student's characterisation (gender, age, academic year, class timetable, class attendance);
2. Student's perception regarding his/her own learning style and learning style used by the student during the development of the PW. The learning styles used were based on the Kolb theory [21];
3. Technical Skills (TS) acquired through PW; 12 items from which 7 are evaluated according to a 5-point Likert scale (1—Very Poor, 2—Poor, 3—Average, 4—Good, 5—Very Good) and the remaining with a "No/Yes" answer;
4. Concept Understanding (CUnd) with PW development, evidenced by a group of sentences given as multiple choice based on 5 sentences;
5. Soft Skills (SS) acquired through the PW development; 5 items evaluated according to a 5-point Likert scale of agreement (1—strongly disagree, 2—disagree, 3—neither agree nor disagree, 4—agree, 5—strongly agree);
6. Course Organisation and Functioning (COF): 15 items, of which 14 are evaluated according to a 5-point Likert scale of agreement (1—strongly disagree, 2—disagree, 3—neither agree nor disagree, 4—agree, 5—strongly agree) and one with multiple choice;
7. Activity Effectiveness (AE), which compares teaching methodologies: 5 items evaluated according to a 5-point Likert scale of agreement (1—lowest effectiveness to 5—highest effectiveness).

2.2.2. Methodology

The questionnaire, in paper format, was handed out to students and, after an explanation about the objectives of the study, answered it on a voluntary basis. The students completed the questionnaire after a moment of individual evaluation of STFLU course, in the 1st semester of 2016/17 and 2017/18 academic years, which took around 10 minutes. A total of 108 students participated in the study. This corresponds to 90.8% of the enrolled students; moreover, this was all the students that attended classes.

2.2.3. Student Characterisation

Table 1 summarises the main student characteristics. Considering the academic years 2016/17 and 2017/18, most of the students (70.4%) are female. The average age was 19.7 years (SD = 1.44, ranged from 19 to 27 years) and the majority of the respondent students (66.7%) were aged 19 years. Fluid Transport Systems is a curricular unit of the 2nd year; however, six students (around 6%) from the 1st year were permitted to attend the course in advance.

Table 1. Descriptive statistics of the student characteristics.

	2016/17	2017/18	Total
Respondent (%)	87.7	93.6	90.8
Gender			
Male (%)	32.0	27.6	29.6
Female (%)	68.0	72.4	70.4
Age			
19 (%)	78.0	56.9	66.7
20– 21 (%)	14.0	32.7	24.1
≥22	8.0	10.4	9.2
Mean Age (x ± SD)	19.5 ± 1.17	19.9± 1.62	19.7 ± 1.44
Regime of Class			
Daytime (%)	98.0	93.0	95.3
After work (%)	2.0	7.0	4.7
First time Attending Students (%)	84.0	89.5	86.9

Students attended the course on different timetables (95.3% in daytime and 4.7% after working hours), and 86.9% of students attended the course for the first time, with a maximum of three times (2.8%). Regarding PW, 3.7% of all the students developed this work individually, 20.8% in pairs, 72.6% in groups of three elements and 5.7% in groups of four elements. 49.1% of the students held a PW of this kind for the first time.

3. Results and Discussion

To understand the changes made in the assessment methodology of STFLU, it is important first to analyse the attendance and grades obtained during the last years (Section 3.1). The students' perceptions with respect to the impact that changes in assessment have in the acquisition and understanding of concepts are presented in Section 3.2.

3.1. Grades and Attendance

The average final grade for those students who appeared at least at one exam moment, the average final grade and the percentage of students who were approved in the STFLU course, are shown in Figure 2 for the different academic years (from 2006/07 to 2017/18).

According to Figure 2, the lowest approval percentage of students (about 70%) occurred in the academic year 2009/10, the exact year in which PW was not done. The average student grade based on approvals did not change greatly during that period and was around 12 out of 20. It would be premature, based on this amount of data, to correlate this value with the absence of PW in 2009/10. However, for the two last academic years (2016/17 and 2017/18), it is evident that there is a similarity of the average grade based on approvals when compared with the average grade based on students that came at least to one exam. Therefore, almost all students who presented themselves to an evaluation had a positive grade. This can also be reinforced by the analysis of Figure 3.

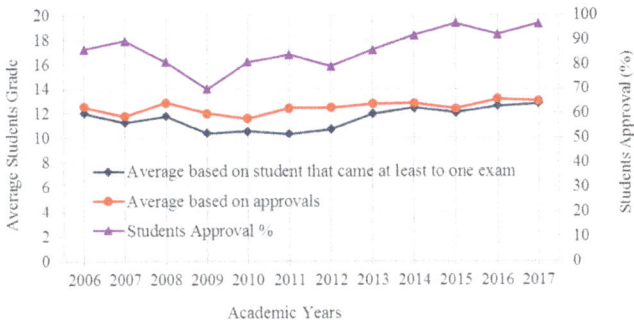

Figure 2. Average of the final grades of students and the percentage of students who were approved for the academic years 2006/07 to 2017/18 (note that the xx axis value corresponds to the beginning of the academic year, i.e., 2006 corresponds to 2006/07).

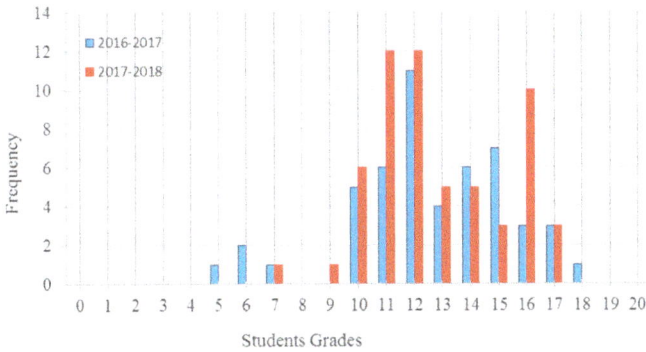

Figure 3. Final grade distribution (number of students that obtained each of the possible grades between 0 and 20) for the academic years 2016/17 and 2017/18.

Figure 3 shows the final grade distribution (number of students that obtained each of the possible grades between 0 and 20) for the academic years 2016/17 and 2017/18. There were few failing grades in these two academic years (grades lower than or equal to 9), corresponding to 3.4%.

Figure 4 shows the average attendance (to theoretical classes) during the academic years between 2013/14 and 2017/18. The attendance data before 2013/14 were not available.

Figure 4. Attendance at theoretical classes for the academic years 2013/14 to 2017/18 (note that the xx axis value corresponds to the beginning of the academic year, i.e., 2013 corresponds to 2013/14).

As attendance at theoretical-practical classes (TP) is mandatory, and the student can only miss 1/3 of the classes (about 5 in 15 classes), attendance is high. According to Figure 4, the level of attendance at theoretical classes was only higher than fifty percent in the last two academic years, which coincides with the period in which more changes were made to the assessment methodology. In that period, six more points of assessment were introduced.

Figure 5 shows the attendance for each theoretical class during the academic years 2016/17 and 2017/18 in more detail.

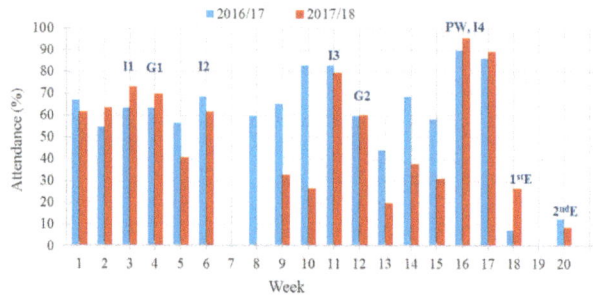

Figure 5. Attendance for each theoretical class throughout the academic year 2016/17 and 2017/18. Ii: Individual moment i, Gi: Group moment i, PW: Practical Work, E: Exam (individual, normal and supplementary).

It is evident that the highest attendance at theoretical classes corresponds to the evaluation days (see the moments identified in Figure 5). Strangely, students missed classes during the week before the assessment moment (see moment I2). However, as stated by Gurung [22], some students tend to do what they should not do, namely, skip classes. Moreover, this behaviour appears not to have had a negative effect on the students' performance, since the majority of them passed the attendance requirement. In addition, this can be confirmed by the low attendance at moments 1st E (normal exam) and 2nd E (supplementary exam).

Comparing year 2016/17 to 2017/18, a higher attendance in the 1st E (1st Exam) in the year 2017/18 can be seen, because a minimum frequency of attendance was instituted. The low attendance at theoretical classes that have no moments of evaluation in 2017/18 when compared to 2016/17 seems to be a characteristic of the year. When comparing the attendance of all courses of the 2nd year/1st semester (Figure 6), it is obvious that low attendance was a common factor among all of them. This difference observed between 2016/17 and 2017/18 seemed not to influence either the average attendance (Figure 4) or the students' mean final grade (Figure 2). Based on these numbers, and as confirmed in the study by Alhija [23], students perceived assessment as the most important of the five teaching dimensions (goals to be achieved, relations with students, teaching methods and characteristics, assessment, long-term student development); therefore, they tended not to miss these moments. The teaching methods were of as little importance as the goals to be achieved and the relations with students.

So the assessment methodology changed the students' attendance at theoretical classes significantly. In general, the majority of the students only attended theoretical classes when they were forced to do it!

This, in a certain way, is in line with the thoughts of Gurung [22] with respect to the idea that teachers can modify the method to assess different levels of classes, and that they should help students to prepare more effectively for exams or other types of assessment. In addition, they can also inform students about study techniques that may be more beneficial to them, since the most important thing is how students study.

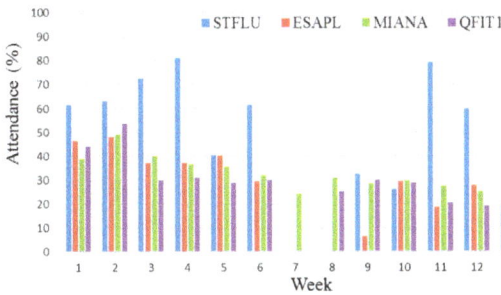

Figure 6. Attendance for each theoretical class of the 2nd year/1st semester courses throughout the academic year 2017/18 (ESAPL—Statistical Methods; MIANA—Methods of Instrumental Analysis; QFIT1—Chemical Physics and Thermodynamics).

3.2. Student Perceptions

The results analysis of the students' perceptions regarding the STFLU course are presented and discussed below. The SPSS statistical tool was used for data analysis [21].

As the data collected from the students do not follow normality (normality was checked with the Shapiro-Wilk test for normality of the data), non-parametric tests were considered in the analysis and to be able to provide answers to the research questions raised (Mann-Whitney (U), for the comparison of two independent samples means as an alternative to the independent sample t-test, and the Spearman's correlation coefficient (r_S) to study and measure the strength of the relationship between two ranked variables/items). A significance level of 5% was considered, meaning that differences were considered to be statistically significant for $p < 0.05$. Both tables and figures are used, when pertinent, to summarise information and to illustrate specific trends or features.

The following research questions will be answered in the subsequent sections:

RQ—How do students perceive the assessment methodology?

sRQ1—How important is the PW in the students' learning process?

sRQ2—Is the PW perceived as an autonomous tool?

sRQ3—Is the PW effective at preparing students for the challenges of professional life?

sRQ4—Which competences did students acquire with the PW?

sRQ5—Are the students confident with the knowledge acquired with the PW to size a proper fluid transport system?

3.2.1. How Do Students Perceive the Assessment Methodology? (RQ)

To answer RQ, from the 15 items evaluated in part 6 of the questionnaire (course organisation and functioning, COF), 14 will be analysed (COF1 to COF14). Furthermore, the individual (Ii) and final students' grades will be related with some of these items allowing to understand possible relations between both. The 14 items under analysis are:

COF1: The examples used by the teachers help the understanding of the contents.

COF2: The teachers try to contextualise the contents in a professional perspective.

COF3: I am interested in learning the contents of the CU.

COF4: I participate in discussions in the classroom.

COF5: The quantity of contents covered in the CU is adequate.

COF6: The practical works/exercises indicated are appropriate to the contents covered in the CU.

COF7: In general, the CU meets my expectations.

COF8: The CU is essentially theoretical.

COF9: Theory is important to understand the practical applications.

COF10: The practical applications proposed are sufficient to understand the concepts.

COF11: The practical applications proposed are sufficient to understand the real applicability of the concepts.

COF12: The evaluation through questions/problems and small tests during theoretical lessons was beneficial to my learning.

COF13: For the intermediate assessment, I prefer several questions/problems and small tests during theoretical classes.

COF14: For the intermediate assessment, I prefer to carry out a single moment of evaluation/test.

Despite a general downward trend, on average, from the 2016/17 academic year to 2017/18, the students' evaluations in these COF items show similar behaviour (Table 2), i.e., there was no statistically significant change in the students' evaluation between the two academic years (for all the items under analyses $p > 0.05$). Since no significant differences were obtained, from here on, the results and analysis will be discussed as a whole.

Table 2. Statistical summary for the considered COF items.

Item	Year	N	Min	Max	Mean	Std. Deviation	Median	Statistics U
COF1	2017	50	3	5	4.40	0.573	4	1153.0
	2018	56	2	5	4.18	0.636	4	
COF2	2017	50	3	5	4.28	0.573	4	1196.5
	2018	56	2	5	4.09	0.640	4	
COF3	2017	50	3	5	4.20	0.535	4	1244.0
	2018	56	2	5	4.04	0.713	4	
COF4	2017	50	2	5	3.44	0.884	3	1147.5
	2018	56	2	5	3.13	0.875	3	
COF5	2017	50	2	5	4.06	0.712	4	1376.5
	2018	56	1	5	3.98	0.863	4	
COF6	2017	50	2	5	4.28	0.671	4	1354.5
	2018	56	1	5	4.18	0.876	4	
COF7	2017	50	3	5	4.16	0.584	4	1157.0
	2018	55	1	5	3.91	0.800	4	
COF8	2017	50	1	5	2.40	0.881	2	1383.0
	2018	56	1	4	2.38	0.822	2	
COF9	2017	48	1	5	3.77	0.778	4	1301.5
	2018	56	1	5	3.68	0.936	4	
COF10	2017	50	2	5	3.82	0.850	4	1238.0
	2018	53	1	5	3.75	0.731	4	
COF11	2017	50	2	5	3.74	0.899	4	1398.0
	2018	54	1	5	3.78	0.691	4	
COF12	2017	50	1	5	4.02	1.020	4	1398.0
	2018	56	1	5	4.05	0.942	4	
COF13	2017	50	1	5	3.82	1.273	4	1320.5
	2018	56	1	5	4.02	1.036	4	
COF14	2017	50	1	5	2.18	1.320	2	1370.0
	2018	56	1	5	2.23	1.348	2	

The lowest result obtained was for items COF14 (For the intermediate assessment, I prefer to carry out a single moment of evaluation and test) and COF8 (The CU is essentially theoretical), with average values lower than 3, showing a disagreement in opinion regarding these two items. However, this behaviour in itself suggests a positive perception regarding the change in the assessment of the theoretical topics such as Fluid Mechanics, making them more practical.

Conversely, the COF items with the highest mean (values higher than 4, showing an agreement in opinion) was COF1 (The examples used by the teachers help the understanding of the contents), COF6 (The practical work/exercises indicated are appropriate to the contents covered in the CU), COF2 (The teachers try to contextualise the contents in a professional perspective), COF3 (I am interested in learning the contents of the CU), COF7 (In general, the CU meets my expectations), COF5 (The quantity of contents covered in the CU is adequate) and COF12 (The evaluation through questions/problems and small tests during theoretical lesson was beneficial to my learning). COF13 (For the intermediate assessment, I prefer several questions/problems and small tests during theoretical classes) presents similar results to COF12, although slightly lower. This could indicate that there is a difference between what students prefer and what is beneficial in terms of learning for them. These results are to some degree in line with the results obtained by Struyven and co-authors in their review work [24], in that what students perceive regarding assessment seems to have a considerable impact on students' learning approaches, and vice versa.

To understand if there is a relationship between the individual (Ii) and students' final grades with respect to the perception that students have regarding assessment methodology, some of the COF items (COF12 and COF14) will be used. Figure 7 shows the average of the students' grades obtained in the four individual assessment moments against the students' perceptions regarding items COF12 and COF14. The individual assessment is the core of the evaluation, and accounts for 83.5% of the final grade (Figure 1).

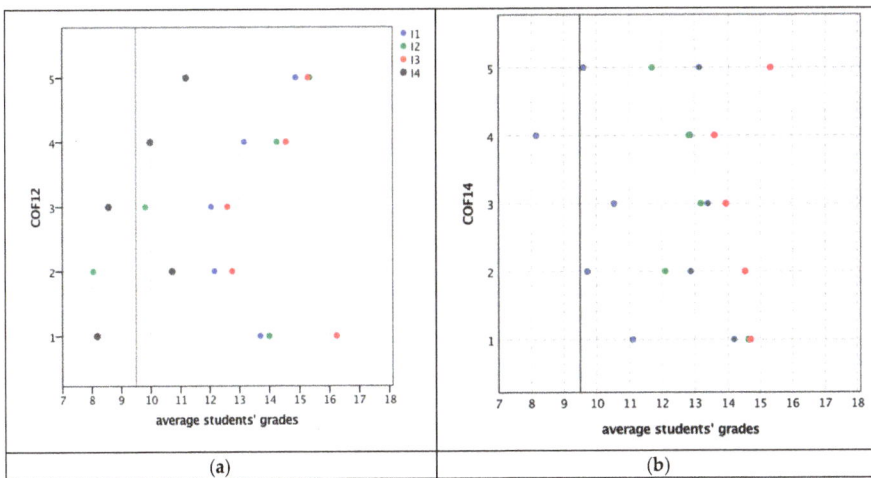

Figure 7. Average student grades versus (**a**) COF12 student evaluation and (**b**) COF14 student evaluation.

As can be seen, there is no defined pattern between the grades in different moments and agreement level with the COF12 item statement. For example, students that are strongly in disagreement (1) with "The evaluation through questions/problems and small tests during theoretical lessons was beneficial to my learning" obtained the highest and the lowest grades (on average). However, all the students that are in complete agreement (5) obtained a positive grade (average ranged from 11 to around 15). Regarding COF14, "For the intermediate assessment, I prefer to carry out a single moment of evaluation and test", the behaviour obtained was more dispersed, making the interpretation more difficult (Figure 7b). That is, for the same individual grade (Ii), students show different evaluations.

By considering the students' final grades, the distribution according to the corresponding students' items evaluation undergoes some changes (Figure 8). Each mark size and colour intensity shown in Figure 8 reflects the number of students in each grid cell (the darkest and largest mark corresponds to 9 students, and then decreases successively until 1). Notice that the levels of agreement for COF12 (Figure 8a) for the 5.6% students that failed (final grade $\leq$ 9.5) were somewhat positive (3—neither agree nor disagree, 4—agree, 5—strongly agree). It is important, however, to emphasise that 72.6% of students with final grades $\geq$ 9.5 agree and strongly agree that the evaluation adopted was beneficial to their learning. Figure 8b shows an opposite evaluation behaviour when looking to the opinions regarding intermediate assessment, where the majority of students (1—strongly disagree, 2—disagree) do not prefer to carry out a single moment of evaluation and test (COF14) independently of the final grade.

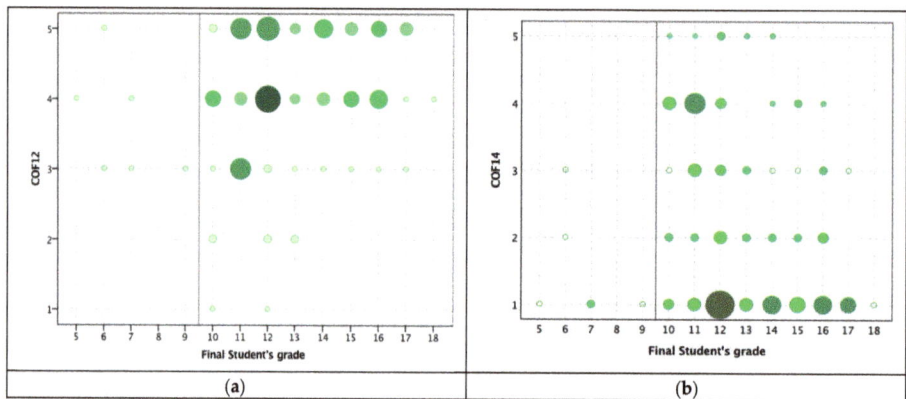

Figure 8. Students' final grades versus (**a**) COF12 student evaluation and (**b**) COF14 student evaluation (marker size and colour intensity reflect the number of students in each grid cell).

3.2.2. How Important is PW in the Students' Learning Process? (sRQ1)

As a measure of how important students perceived the PW to be in their learning process, a set of technical skills acquired with the PW (part 3 of the questionnaire, TS items) was considered:

TS4: PW is a useful tool in supporting the CU,

TS6a: Overall, the PW helped me to assimilate the concepts transmitted throughout the semester,

TS6b: Overall, the PW has made my learning more objective,

TS6d: Overall, the PW motivated me to the CU,

TS7: Overall, I felt motivated to carry out the PW,

TS9: The PW motivated me to learn the CU contents,

TS10: Would you recommend doing a PW as a teaching/learning activity?

The two last items (TS9 and TS10) are "No/Yes" questions.

Table 3 and Figure 9 summarise and illustrate the distribution of the obtained data, respectively.

Largely, students agreed (Figure 9, more than 50% of agreement, 4 and 5) that the PW is an important and useful tool in supporting the CU (TS4). This item is the one with the highest evaluation (Table 3, mean = 4.34). Even the item with the lowest evaluation (TS6d, mean = 3.80) reflects an agreement. In fact, this is the only item where the evaluation of 1 (strong disagreement) was obtained. The percentage of disagreement (1 and 2) corresponds to 3.7%, meaning that these students do not consider that the PW development has a motivating effect on the CU.

Despite the previous results, a majority of students (90.7%) consider that the PW motivated them to learn the CU contents (TS9), recommending (96.3%) the PW as a teaching activity in their learning process (TS10).

Knight, in their work [25], stated that, related to group work development, students give more importance to students' overall 'group experience' than the related learning outcomes/skills developed in group work. This could explain why some students did not consider the PW to be motivating for the CU.

Table 3. Statistical summary for the considered Technical Skill (TS) items.

Item	N	Min.	Max.	Mean	Std. Deviation	Median
TS4	108	2	5	4.34	0.713	4
TS6a	108	2	5	4.13	0.698	4
TS6b	108	2	5	4.12	0.680	4
TS6d	108	1	5	3.80	0.783	4
TS7	108	2	5	3.91	0.803	4
TS9	108			90.7% Yes		
TS10	108			96.3% Yes		

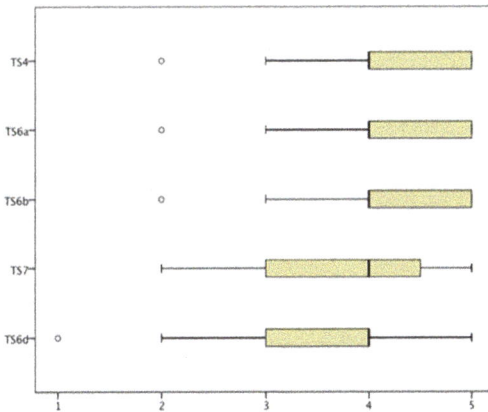

Figure 9. Students' evaluation for TS Technical Skills.

3.2.3. Is the PW Perceived as an Autonomous Tool? (sRQ2)

To understand how students perceived the PW as an autonomous tool, item TS12, with a "No/Yes" answer (Do you consider that the PW should be less supervised by the teacher?), belonging to part 3 of the questionnaire (Technical Skills acquired with the PW), will be analysed. Since the students were asked to justify their answers to this question with words, this will also be analysed.

The majority of the students (94.4%) answered "No". The justifications were related to the support that the teacher must give to students. For example: "it is necessary to have indications from the teacher for the development of the PW"; "teacher' support is essential"; "in case of less support, we (students) would not manage to develop the PW". These last comments are in opposition to the comments given by the minority of students that answered "Yes" (5.6%): "the PW obliges students to look for information".

As previously mentioned (Section 3.1), students' attendance at theoretical classes was relatively low (high attendance only occurred at evaluation moments, as expected). It is in these classes that important and necessary PW information is given, so if students were present in these classes, the need for indications by the teacher could be diminished. Then, the teacher's support would only be necessary for some doubts. As emphasised by one student, "even though the contents are taught, there are always some aspects that are specific and vary".

Answers like "if the PW was less supervised the students would fill lost and would neglect it" shows that some students were not able to perceive the PW as a tool for applying their knowledge and understanding in solving new and practical problems.

It seems that, independently of approaches to learning and to teaching, student motivation is mainly associated with intrinsic or autonomous motivation and each student's learning profile; moreover, student-centred learning environments do not guarantee success at motivating students to learn [26].

3.2.4. Is the PW Effective for Preparing Students for the Challenges of Professional Life? (sRQ3)

To understand how students perceived the PW and their level of agreement regarding PW effectiveness as an autonomous tool, item TS11 (Do you consider that the PW should be carried out before the contents have been taught?) was related to the five items that measure the activity effectiveness (AE):

AE1—theoretical class where the teacher lectures the contents and a theoretical-practical class where applied exercises are solved,

AE2—previous study of theoretical concepts proposed by the teacher and discussion in theoretical class of the application of these concepts,

AE3—students individually solve a global question in the theoretical class using the knowledge they have,

AE4—theoretical class where the teacher presents the contents using practical examples and a theoretical-practical class where applied exercises are solved,

AE5—students solve a problem in groups in the theoretical class using the knowledge they have.

Table 4 presents the corresponding statistics in terms of minimum, maximum, mean and median values and the standard deviation.

Table 4. Statistical summary for the considered Activity Effectiveness (AE) items.

item	n	Min.	Max.	Mean	Sth. Deviation	Median
AE1	106	2	5	4.10	0.675	4
AE2	106	1	5	3.69	0.919	4
AE3	105	1	5	3.76	1.033	4
AE4	106	1	5	4.30	0.783	4
AE5	106	2	5	3.96	0.827	4

On average, students considered the traditional teaching/learning methodology to be the most effective, that is, a theoretical class where the teacher presents the contents using practical examples and a theoretical-practical class in which applied exercises are solved (AE4, with 4.30 on a scale of 5), and where in the theoretical class, the teacher presents the contents and a theoretical-practical class in which applied exercises are solved (AE1, with 4.10 in a scale of 5).

Students considered the methodology in which students are required to undertake previous study of theoretical concepts proposed by the teacher, followed by a discussion in the theoretical class on the application of these concepts, to be of lower effectiveness (AE2, with 3.69 in a scale of 5).

To understand students' choices regarding the effectiveness of the PW as an autonomous tool, these 5 items were correlated with item TS11. Table 5 summarises the values of the Spearman coefficients obtained. None of the 5 items (AE1, AE2, AE3, AE4 and AE5) show a significant correlation with item TS11. To some extent, these results confirm earlier findings. However, slightly increasing the significance level to 6%, two positive correlations could be observed. That is, students who agree that the PW should be carried out before the contents have been taught also agree that they could first study the theoretical concepts proposed by the teacher and then discuss them later in theoretical class (AE2), and also agree that students could solve a problem in groups using the knowledge they have (AE5), understanding and identifying their difficulties.

Table 5. Spearman's correlation coefficient, r_S.

	TS11	AE1	AE2	AE3	AE4	AE5
TS11	1.0					
	.					
AE1	0.025	1.0				
	0.797	.				
AE2	0.186 ‡	0.175	1.0			
	0.056	0.072	.			
AE3	0.082	0.120	0.363 **	1.0		
	0.406	0.223	0.000	.		
AE4	−0.101	0.275 **	0.122	0.283 **	1.0	
	0.301	0.004	0.212	0.003	.	
AE5	0.185 ‡	0.012	0.315 **	0.514 **	0.192 *	1.0
	0.058	0.905	0.001	0.000	0.048	.

** Correlation is significant at the 0.01 level (2-tailed). * Correlation is significant at the 0.05 level (2-tailed). ‡ Correlation is significant at the 0.06 level (2-tailed).

3.2.5. Which Competences Did Students Acquire with the PW? (sRQ4)

To answer sRQ4, the five items in the soft skills part of the questionnaire (part 5) will be analysed concerning what the PW allowed:

SS1: to encourage collaborative work,
SS2: to stimulate my intellectual curiosity,
SS3: to provide necessary knowledge for my area of study,
SS4: to relate this CU to other CUs,
SS5: to apply the acquired concepts during the PW development in different CUs.

As for the previous items, for this set also, no statistical differences were obtained between the two academic years with regard to the way students evaluate them (Table 6).

Table 6. Statistical summary for the three SS items.

Item	Year	N	Min	Max	Mean	Std. Deviation	Median	Statistics U
SS1	2017	50	1	5	4.02	0.845	4	1391.0
	2018	56	3	5	4.09	0.695	4	
SS2	2017	51	3	5	3.98	0.678	4	1413.0
	2018	56	2	5	3.98	0.751	4	
SS3	2017	51	2	5	4.08	0.659	4	1415.5
	2018	56	2	5	4.09	0.668	4	
SS4	2017	51	1	5	3.37	0.937	3	1419.0
	2018	56	1	5	3.45	0.933	3	
SS5	2017	51	1	5	3.41	0.898	4	1374.0
	2018	56	1	5	3.50	1.027	3	

Figure 10 shows the distribution of the students' evaluation for the five SS items considered in the questionnaire. The items evaluated more positively corresponded to SS3 (Provide necessary knowledge for my area of study), SS1 (Encourage collaborative work) and SS2 (Stimulate my intellectual curiosity).

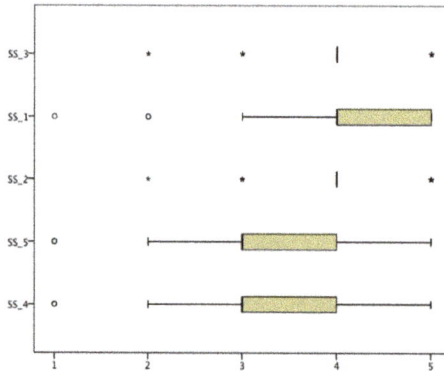

Figure 10. Students' evaluation for SS Soft Skills.

The students were able to identify that the development of the PW helped in providing knowledge of the area under study (SS3). Only two students stated disagreement. Regarding SS4 and SS5, the majority of students still experienced difficulties in relating the acquired concepts to other subjects (55.1% and 51.4%, respectively, for disagree and do not know). This is not surprising, since the courses until the first semester of the second year are mainly basics courses (mathematics, physics and chemistry). Although they have an integrating laboratory (LABO3) in which they practice Fluid Mechanics simultaneously with STFLU, they do not relate STFLU to this other CU.

These results, together with the results obtained in Section 3.2.1 (COF12), reinforce the idea that the PW and several moments of assessment can be used as tools promoting the acquisition of knowledge in Fluid Mechanics.

3.2.6. Are the Students Confident with the Knowledge Acquired with the PW in Sizing a Proper Fluid Transport System? (sRQ5)

To understand how confident students are with the knowledge acquired with the PW to size a proper fluid transport system, group 4, Concepts' Understanding (CUnd) with PW development, was analysed. To this end, five sentences were given, and the student selected the most appropriate situation:

CUnd_a: with the PW development, now I am able to select a pump and size the involved system (pump and system characteristics, costs ...) for any situation,

CUnd_b: with the PW development, now I am able to select a pump and size the involved system (pump and system characteristics, costs ...) for a house/building,

CUnd_c: with the PW development, now I am able to select a pump (pump characteristics, costs ...) for any situation,

CUnd_d: with the PW development, now I am able to select a pump (pump characteristics, costs ...) for a house/building,

CUnd_e: with the PW development, I am still not able to select a pump and size the involved system (pump and system characteristics, costs ...) for a house/building.

Figure 11 shows the percentage of students' answers for each of the CUnd items.

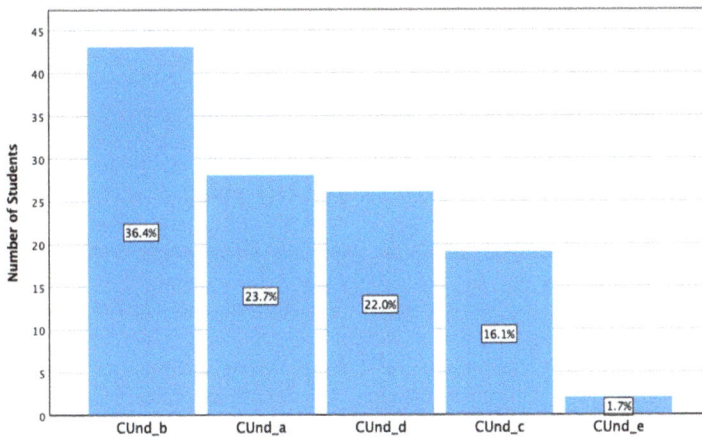

Figure 11. Students' answers for CUnd with PW development.

As can be seen, the majority of students (60.1 %) were confident when selecting a pump and sizing the involved system (pump and system characteristics, costs ...). Some students (38.1 %) were only confident when selecting a pump (pump characteristics, costs ...) without sizing the complete flow system. Only 1.7 % of students (corresponding to 2 students) did not feel confident when sizing this type of system. These results reflect the problem associated with group work; that is, some groups split the tasks among them and only become aware of the contents that they developed, forgetting to integrate all of the contents [27].

Nevertheless, these results show that students are in general confident with the knowledge acquired through PW.

4. Conclusions

The main objective of this study was to evaluate students' perceptions regarding different types of assessment and to analyse the impact of the changes in the assessment methodology on attendance and on the students' final grades.

It was verified that the assessment methodology changed the students' attendance at theoretical classes significantly. In general, the majority of the students only attended theoretical classes when they were forced to do it! The introduction of several assessment moments increased students' attendance at the theoretical classes, mainly on the days on which assessments occurred. In fact, the absence at theoretical classes was somewhat problematic, and the methodology enhanced the attendance at six out of fifteen theoretical classes. This assessment methodology allows the students to be focused on and motivated by the learning process. Consequently, the majority of students who presented themselves to an evaluation passed the attendance requirement.

The development of the PW as an evaluation method has positive repercussions for the way students focus on the course contents and keep up with the subjects taught.

Studying throughout the semester (as opposed to studying only for a final exam) facilitates good knowledge acquisition and allows a better success with respect to final grades and also to the preparation for future professional life.

In general, students consider the contents, practical examples and contextualisation of the CU to be adequate and positive with respect to their learning process and preparation for professional life.

Students prefer assessment through several questions/problems and small tests during the theoretical lessons, and consider it more beneficial for their learning, rather than carrying out a single evaluation and test. Even students that fail (final grade ≤ 9.5) classified the assessment methodology based on several questions/problems and small tests during theoretical lessons positively (level of agreement between 3 and 5).

Largely, students agree that the PW is an important tool in their learning process that helps them to assimilate concepts and make their learning more objective. The majority of students recommend the PW as a teaching activity.

On average, students considered the traditional teaching/learning methodology (a theoretical class, where the teacher presents the contents using practical examples, and a theoretical-practical class, where applied exercises are solved) to be the most effective. In line with this, students did not agree that the PW should be carried out before the contents had been taught; they did not intend to use the PW as an autonomous tool.

In general, students were confident with the knowledge acquired with the PW and felt able to size fluid transport systems.

The majority of students were able to identify that the development of the PW and several moments of assessment helped in providing knowledge on the area under study, encouraged their collaborative work and stimulated their intellectual curiosity, but they still experienced difficulties in relating the acquired concepts with other subjects.

The authors believe that the students acquired new competences, in particular autonomous work, that may be applied in other courses and in their professional life.

Author Contributions: All authors conceived and designed the experiment. T.S.-E. and C.M. collected data. C.P.L. analysed the data with the SPSS statistical tool. All authors contributed to the final manuscript.

Funding: This research received no external funding.

Acknowledgments: The authors would like to express their acknowledgment to all students who agreed to collaborate in this study. The authors also thank the Research Centre CIETI (Centro de Inovação em Engenharia e Tecnologia Industrial) and FCT—Fundação para a Ciência e Tecnologia for all the support provided in the scope of the projects COMPETE: POCI-01-0145-FEDER-007043, UID/CEC/00319-2019 and UID-EQU-04730-2019.

Conflicts of Interest: The authors declare no conflict of interest.

References

1.   Cansino, J.M.; Román, R.; Expósito, A. Does Student Proactivity Guarantee Positive Academic Results? *Educ. Sci.* **2018**, *8*, 62. [CrossRef]
2.   Lukkarinen, A.; Koivukangasa, P.; Seppäläa, T. Relationship between class attendance and student performance. *Procedia Soc. Behav. Sci.* **2016**, *228*, 341–347. [CrossRef]
3.   Ma, X.; Luo, Y.; Zhang, L.; Wang, J.; Liang, Y.; Yu, H.; Wu, Y.; Tan, J.; Cao, M. A Trial and Perceptions Assessment of APP-Based Flipped Classroom Teaching Model for Medical Students in Learning Immunology in China. *Educ. Sci.* **2018**, *8*, 45. [CrossRef]
4.   Billett, S. Learning through Practice. In *Learning Through Practice. Models, Traditions, Orientations and Approaches*, 1st ed.; Billett, S., Ed.; Springer Science + Business Media B.V: Brisbane, Australia, 2010; pp. 1–20.
5.   Lu, H.; Zhang, X.; Jiang, D.; Zhao, Z.; Wang, J.; Liu, J. Several Proposals to Improve the Teaching Effect of Fluid Mechanics. In *Emerging Computation and Information teChnologies for Education*; Mao, E., Xu, L., Tian, W., Eds.; Springer: Berlin/Heidelberg, Germany, 2012; pp. 187–191.
6.   Hunsu, N.; Abdul, B.; Van Wie, B.J.; Brown, G.R. Exploring Students' Perceptions of an Innovative Active Learning Paradigm in a Fluid Mechanics and Heat Transfer Course. *Int. J. Eng. Educ.* **2015**, *31*, 1200–1213.
7.   Absi, R.; Nalpace, C.; Dufour, F.; Huet, D.; Bennacer, R.; Absi, T. Teaching Fluid Mechanics for Undergraduate Students in Applied Industrial Biology: From Theory to Atypical Experiments. *Int. J. Eng. Educ.* **2011**, *27*, 550–558.
8.   Sorensen, E. Implementation and student perceptions of e-assessment in a Chemical Engineering module. *Eur. J. Eng. Educ.* **2013**, *38*, 172–185. [CrossRef]
9.   Rahman, A. A blended learning approach to teach fluid mechanics in engineering. *Eur. J. Eng. Educ.* **2017**, *42*, 252–259. [CrossRef]
10.   Gynnild, V.; Myrhaug, D.; Pettersen, B. Introducing innovative approaches to learning in fluid mechanics: A case study. *Eur. J. Eng. Educ.* **2007**, *32*, 503–516. [CrossRef]

11. Golter, P.B.; Thiessen, D.B.; Van Wie, B.J.; Brown, G.R. Adoption of a non-lecture pedagogy in chemical engineering: Insights gained from observing an adopter. *J. STEM Educ.* **2012**, *13*, 52–62.

12. Prince, M.J.; Felder, R.M. Inductive Teaching and Learning Methods: Definitions, Comparisons, and Research Bases. *J. Eng. Educ.* **2006**, *95*, 123–138. [CrossRef]

13. Boylan-Ashraf, P.C.; Freeman, S.A.; Shelley, M.C.; Keleş, Ö. Can Students Flourish in Engineering Classrooms? *J. STEM Educ.* **2017**, *18*, 16–24.

14. Schmidt, H.G.; Rotgans, J.I.; Yew, E.H. The process of problem-based learning: What works and why. *Med. Educ.* **2011**, *45*, 792–806. [CrossRef] [PubMed]

15. Yadav, A.; Subedi, D.; Lunderberg, M.; Bunting, C.F. Problem-based Learning: Influence on Students' Learning in an Electrical Engineering Course. *J. Eng. Educ.* **2011**, *100*, 253–280. [CrossRef]

16. Daly, S.R.; Mosyjowski, E.A.; Seifert, C.M. Teaching Creativity in Engineering Courses. *J. Eng. Educ.* **2014**, *103*, 417–449. [CrossRef]

17. Gibbs, G.; Simpson, C. Conditions under which Assessment supports Student Learning. *Learn. Teach. High. Educ.* **2004**, *1*, 3–31.

18. Hettiarachchi, E.; Huertas, M.A.; Mor, E. E-Assessment System for Skill and Knowledge Assessment in Computer Engineering Education. *Int. J. Eng. Educ.* **2015**, *31*, 529–540.

19. Soares, F.; Leão, C.P.; Guedes, A.; Brás-Pereira, I.; Morais, C.; Sena-Esteves, T. Interpreting students' perceptions in fluid mechanics learning outcomes. *Educ. Knowl. Soc.* **2015**, *16*, 73–90. [CrossRef]

20. Sena-Esteves, T.; Morais, C.; Guedes, A.; Brás-Pereira, I.; Ribeiro, M.M.; Soares, F.; Leão, C.P. Teaching Impact and Evaluation Methodology Assessment in a Fluid Mechanics Course: Student's Perceptions. In Proceedings of the ASME 2017 International Mechanical Engineering Congress and Exposition (ASME IMECE2017), Tampa, FL, USA, 3–9 November 2017.

21. Field, A. *Discovering Statistics Using SPSS*; SAGE, Publications Ltd.: London, UK, 2009.

22. Gurung, R.A.R. How do students really study (and does is matter)? *Teach. Psychol.* **2002**, *2*, 149–155.

23. Alhija, F.N. Teaching in higher education: Good teaching through students' lens. *Stud. Educ. Eval.* **2017**, *54*, 4–12. [CrossRef]

24. Struyven, K.; Dochy, F.; Janssens, S. Students' perceptions about evaluation and assessment in higher education: A review. *Assess. Eval. High. Educ.* **2005**, *30*, 325–341. [CrossRef]

25. Knight, J. Comparison of student perception and performance in individual and group assessments in practical classes. *J. Geogr. High. Educ.* **2004**, *28*, 63–81. [CrossRef]

26. Baeten, M.; Kyndt, E.; Struyven, K.; Dochy, F. Using student-centred learning environments to stimulate deep approaches to learning: Factors encouraging or discouraging their effectiveness. *Educ. Res. Rev.* **2010**, *5*, 243–260. [CrossRef]

27. Cohen, E.G. Restructuring the Classroom: Conditions for Productive Small Groups. *Rev. Educ. Res.* **1994**, *64*, 1–35. [CrossRef]

© 2019 by the authors. Licensee MDPI, Basel, Switzerland. This article is an open access article distributed under the terms and conditions of the Creative Commons Attribution (CC BY) license (http://creativecommons.org/licenses/by/4.0/).

education sciences

MDPI

Article

Eco-design and Eco-efficiency Competencies Development in Engineering and Design Students

Victor Neto

Center for Mechanical Technology and Automation, Department of Mechanical Engineering, University of Aveiro, 3810-193 Aveiro, Portugal; vneto@ua.pt

Received: 30 April 2019; Accepted: 5 June 2019; Published: 7 June 2019

check for updates

Abstract: The development of vital competencies and a mindset to rethink products, production, and business models in engineering and design students is presently of great importance. These future professionals will play a key role in the development of sustainable products. Within Eco-design and Eco-efficiency curricular unit, an assignment was developed that consisted of the development of an eco-design and eco-efficiency study of a given product, provided by a real industrial company. In this paper, the challenge description and application are reported, as well as the key conclusions.

Keywords: circular Economy; eco-design; eco-efficiency; engineering education

1. Introduction

There is a need to produce goods with a lower impact on nature, reducing the use of primary raw materials, minimizing energy consumption, and promoting long and circular product life cycles [1–4]. These are the joint challenges of Industry 4.0 and Circular Economy, that, together, will promote a production and consumption paradigm shift [5–9].

Industry 4.0 is slowly changing the way we produce goods and services in order to achieve greater productivity using less material and fewer energy resources. The "Industry 4.0" philosophy seeks to introduce the technological advances that have been achieved in recent years in the field of sensor and control, computing, and automation processes, creating the conditions for new product development processes and new forms of production, including integration of conventional and advanced manufacturing technologies, such as additive fabrication [5]. However, this philosophy of industrial development must also be based significantly on the re-use of raw materials and making use of eco-design and eco-efficiency strategies [6,7]. The price of raw materials is continuously increasing, because of its growing scarcity in many cases, but also because of the social and environmental costs that the extraction and production of new raw materials entail. It is from the integration of these different perspectives that companies can better embrace the circular economy.

The concept of circular economy aims to respond to the challenges of maintaining life quality, without exterminating humanity and the planet [8]. A circular economy is, in principle, regenerative and restorative. Its goal is to keep timely products, components, and materials at the highest level of utility and value. This means that a product, after its use, is not discarded for a landfill or for incineration. It means that after its lifetime, the product continues its life cycle, to be repaired or transformed, giving rise to a regenerated product or to raw materials that will constitute a new product. With this, we keep these products, components, and materials in a closed circle of economic utility, without increasing exponentially the need for new raw materials, nor waste of materials for landfills [9].

Circular economy is perceived to induce regenerative industrial transformations that will pave the way to achieve sustainable production and consumption. The ambition is that the evolution of circular economy based industrial production will have a positive impact on the environment and will also contribute to economic growth [10]. As society increasingly seeks the bases of sustainable living, we are becoming more aware of the key responsibilities that consumers and organizations have. A spotlight shines on company behavior and reveals the importance of encouraging firms to use their resources as efficiently as they can. Nevertheless, companies may lack the information, confidence, and capacity to move to a circular economy, due to a lack of indicators and targets, awareness of alternative circular options, and economic benefits, and, especially, the existence of skill gaps in the workforce and lack of circular economy related curricula [11]. The development of vital competencies and a mindset to rethink products and production settled in these new concepts in mechanical engineering and product design students is of fundamental importance. It is mandatory that these future professionals play a key role in the development of sustainable products.

There are many initiatives underway to implement the concept, especially in legislative and governmental bodies, NGOs, and consultancy firms, but real practical established approaches are still under construction [8] and need to be supported by examples of new business models [7] that have these concepts in their base. As an example, imagine diamond cutters for cutting glass or ceramic material. The conventional business model goes through a manufacturer, possibly in China, to produce the metal milling cutters, and the same manufacturer, or another, will apply the coating of synthetic diamond microcrystals. Once the production is finished, they will be sold to a company that will use them until the diamonds' cutting efficiency decreases considerably. The final destination of these cutters is the garbage or, at best, a sale for the recovery of the metal. However, these cutters could be re-coated with diamond multiple times, always having the same cutting efficiency as the completely new cutters. With this business model, the material and energy expended to manufacture the metal part would be spared. Despeisse et al. [12] points that the characteristics of additive manufacturing align well with sustainability and circularity principles and hold significant promise for moving society in a more sustainable direction, as these characteristics can be used for repair and remanufacturing and the production of printing filaments, including the commercialization of filament that contains recycled materials, and recycling systems for creating filament.

The literature points out that the well-established eco-design guided by the life cycle assessment of a product and eco-efficient production are vital for the transition to a circular economy [13–19]. Eco-efficiency is based on the idea that fewer natural resources should be used to generate the same, or a greater, amount of economic activity. Whatever the setting, objectives are loosely grouped around sustainability. Eco-efficiency can be seen as a tool for sustainability analysis and development [18]. Eco-design is defined as the integration of environmental aspects into product design and development with the aim of reducing adverse environmental impacts throughout a product's life cycle [19]. It must focus its attention on the phases of the product's life cycle that most significantly affect the environment, so that upon re-designing the product, its environmental impact can be greatly reduced. Therefore, integrating eco-design and eco-efficiency into the product development process can contribute to the development of vital competencies and a mindset to rethink products and production process that will potentially assist the transformation of linear to circular economic business models and can offer several advantages to industry and public organizations, such as economic benefits, legislation fulfilment, innovation and creativity promotion, public image improvement, and employee motivation enhancement.

Although eco-design and eco-efficiency concepts are well-established in both theory and practice, and across a wide variety of contexts, education for sustainable development is quite a challenge, particularly to educate future engineers in this manner, as this type of education demands a departure from the current disciplinary and subject-focused teaching that predominates current educational paradigms, particularly in engineering education [20,21]. Simply integrating conceptual topics into existing courses is not enough, as the current paradigm's approach is too reductionistic to handle multidimensional problems [22,23]. Instead, students must learn to employ system thinking to fully comprehend the challenges. Furthermore, in addition to the key eco-design and eco-efficiency concepts competencies, students must also build an awareness of societal and economic aspects [23].

Different approaches for education and learning have been developed in similar contexts, including gamification methodologies [23–25]. Nonetheless, the integration of teaching and learning in higher education with its surrounding societal reality, in an embeded way, is particularly critical to the development of future professionals with sustainability literacy, as these future professionals will become the agents of change in their workplaces and personal lives [26]. Work-based learning at higher education levels has been highlighted as a pressing need [27–30]. In the current context, it is imperative that students learn how the subjects they address in the classroom are related to the real world [28]. Simultaneously, by putting them in touch with industry, these students are given the possibility to explore career options. Additionally, from this university–student–society interaction, companies have the possibility to interact with potential future employees who will have a better knowledge of the workplace. Higher education institutions, on the other hand, benefit from an increase in student motivation and can improve the relationship between schools and the community. In addition, curricular interaction with society is an ideal methodology to relate the content taught, whether fundamental or applied, to the challenges of the society and achieve a balance between the fundamental technical-scientific competences and the transversal competencies currently required by employers [29].

Having, therefore, the double objective of addressing eco-design and eco-efficiency to promote a circular economy mindset and the integration of teaching and learning with the surrounding industry, the Eco-design and Eco-efficiency (EDEE) curricular unit at the University of Aveiro (Portugal) has been promoting the development of eco-design and eco-efficiency projects for selected products in their industrial production environment, presented by industrial entities of the Aveiro region, to promote an entrepreneurial mindset for the creation of sustainable products and processes, in scope with industrial trends of digitalization and circularity.

In this work, the description of the assignment and its application is reported, as well as the key results obtained. It is relevant to state and analyze the key findings, including if the assignment has contributed to the goal of promoting an entrepreneurial mindset for the creation of sustainable products and processes, as well as promoting a better comprehension of workplace and industry environments.

2. Methods

The Eco-design and Eco-efficiency (EDEE) curricular unit at the University of Aveiro is an optional discipline offered, mainly, to Mechanical Engineering and Product Design and Engineering master students. The case study presented in the present paper has taken place in the academic year of 2018–2019 (semester 1), and the class was composed of 49 students, of which 37 were Mechanical Engineering students, 9 were Product Design and Engineering students, 1 was an Electronic and Telecommunications Engineering master student, and 2 were Mechanical Engineering Erasmus program exchange students (one from Germany and another from Italy). The curricular unit is organized and lectured by one teacher, but several other university teachers and researchers are involved in the connection with the companies.

The main objective of the EDEE course is to promote an entrepreneurial mindset for the creation of sustainable products and processes, in line with industrial trends of digitalization and circularity. Students will be challenged to develop a new product, or rethink a previous product, to decrease resource use intensity, giving priority to the use of renewable materials, including recyclable and/or bio-based materials, and with less hazard and risk (for humans and the environment) and better reuse of materials. To this end, eco-design and eco-efficiency concepts and tools, and their applications in the different stages of engineering and product development, are covered, as well as rules and regulations. Specific issues addressed during the semester include product development and industrial trends; eco-design and eco-efficiency concepts and tools [1–4]; life cycle assessment—ISO 14040 and 14044 standards [31–33]; industrial environmental management systems—ISO 14001:2015 standard [34]; energy management in manufacturing—ISO 50001:2018 standard [35]; lean manufacturing [36,37] and other principles that can assist efficiency production; and product and business design for a circular economy.

The proposed assignment, described in this paper, is very much aligned with the curricular unit objectives. The development of the assignment can be divided into three parts, as illustrated in Figure 1. The first stage is to clearly identify the product and fabrication process (as done by the company). Second, students must perform an eco-design analysis of the product and an eco-efficiency evaluation of the production process. This should be developed as a Life Cycle Assessment (LCA) of the product [31–33], considering the product from cradle to grave. For the LCA analyses, students are encouraged to use Ecolizer (http://ecolizer.be/) or openLCA (http://www.openlca.org/) software. The first is a free online platform that enables a simple comprehensive inventory and environmental impact calculation. The second is an open source and free software for sustainability and life cycle assessment. Although both present limitations for a complete product LCA analysis, they have the necessary requirements for the assignment. Students may calculate the overall environmental impact but also the impact of each phase in the life cycle of a product so that a life cycle phase with a high environmental impact can be tackled or compare the scores of different products with each other.

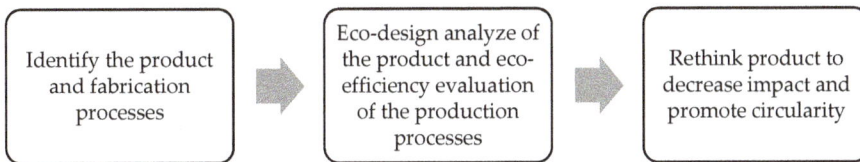

Figure 1. Assignment development stages.

Finally, in the third part, the proposals to decrease the product's impact should be projected. In this last stage, students should rethink the product to decrease the resource use intensity, giving priority to the use of renewable materials, including recyclable and/or bio-based materials, and with less hazard and risk (for humans and the environment) and reuse of materials. "Modularization" of the components, allowing easy disassembly, recovery, reuse, and end-of-life screening (standard components) may be considered, as well as the definition of recycling, reuse, and life-cycle extension criteria, considering possible useful applications of by-products and waste. Groups should pursue ways to attain more efficient and cleaner production models, producing more, at lower prices, with fewer resources, less waste, and less of an impact on the environment. The use of RETScreen (https://www.nrcan.gc.ca/energy/software-tools/7465), a clean energy management software, is given as an example of a tool to use to assist in the pursuit of increased ecoefficiency production. Proposals to convert the business model to a circular economy business model are also encouraged.

The assignment must be developed respecting the pace of the following milestones, distributed homogeneously throughout the semester:

- Milestone 1: Product and production description (current situation). A 3 minute video explaining the product and the production process must be delivered.
- Milestone 2: Life cycle assessment of the product (current situation). A 5 minute video presenting the current LCA of the product and production process must be submitted.
- Milestone 3: Eco-design of the product and production processes eco-efficiency. A 5 minute video explaining the eco-design and eco-efficiency proposals must be handed.
- Milestone 4: Full assignment presentation and discussion, in the company to the company people.
- Milestone 5: Final report delivery. The technical report must have a maximum of six A4 pages complemented with the needed attachments.

Although class materials are all in English, the official language of the curricular unit, the deliverables, presentation, and discussion above mentioned can be in either English or Portuguese. Also, the milestone deliverables can be reformulated at any time until the final exam period.

Because the development of the assignment needs access to production data to properly instruct the inventory stage of the life cycle assessment, students and other people involved in this assignment must compromise to keep all information accessed and deliverables confidential. Complementarily, if the assignment creates truly innovative results that may be the object of intellectual property registration, the copyright will be given to the university and the company. The copyright given to the university will not jeopardize the rights of the students, as well as the teachers and other involved staff, to be designated as creators, inventors, or authors of the invention or creation.

The assignment deliverables were organized by the class teachers on a private webpage. The assignments were developed in groups of 4 to 5 persons, and each group had to be composed of students from at least two different courses. This assignment represented 60% of the final grade of the curricular unit, whether students were in discrete or final evaluation. The additional 40% of the grade was obtained by an individual written test.

For the assignment development, each group selected a product/company from a given pool. This products and companies were arranged by the curricular unit teachers and are presented in Table 1.

Table 1. Companies and products studied in the assignment.

Company	Location	Product(s)
Composite Solutions	Vagos	"Waterlily"
Levira	Oliveira do Bairro	Office desk and cabinet
Mistolim	Vagos	Detergent Pack
Moldit	Loureiro	Gardening vase
OLI	Aveiro	Toilet flush (bathroom)
PNH	Águeda	Restaurant toaster and fryer
Ramalhos	Águeda	Bakery oven

To evaluate the perception and appreciation of the learning and teaching outcomes by the involved agents—the students, corporations, and teachers—simple open-ended questions surveys where developed. The objective was to get information about the assignment contribution to the proposed goals. Is the assignment contributing to the proposed goals? Are students being able to connect the subjects that they learn in the classroom with the challenge case and develop a better comprehension of the workplace? Are students motivated by the assignment? Are companies satisfied by the technical results proposed and with the interaction with students? Is the university's business relationship improved? Are students' fundamental and transversal soft skills improved?

For students, making use of an online form tool, one single open question was placed:

"What is your global appreciation of the assignment? Is this type of evaluation justified in a course of Eco-design and Eco-efficiency? Do you consider that this type of work is relevant to your training? Do you think that initiatives of this kind contribute to a greater alignment between the university and society? What were the most positive and the most negative point of the assignment? What aspects could be improved?"

For companies, again one single open question inquiry was placed and sent by email:

"What is your global appreciation about the assignment that students developed within your company?"

In the case of the teacher and collaborators, the register data was compiled in groups, as part of the assignment self-evaluation routine.

3. Results

The different surveys resulted in 10 responses from students (out of 49 students, representing about 20%) and five responses from companies (out of 7, representing about 70%). The collected data was analyzed and resumed in Tables 2–4 in order to highlight the main ideas.

Table 2. Resume of students' responses to the inquiry.

Students Main Appreciation about the Assignment
"The assignment showed us the difference between an academic assignment and an assignment for a company."
"We realized that some concepts cannot be applied linearly in the company."
"There should be more initiatives such as this one so that we could understand how companies work."
"I enjoyed doing the work. It put us close to the industry and facing real production processes."
"Having a colleague from a different course in the group was very useful since they have complementary experiences and skills."
"It would be good to see some of our solutions being implemented by the company and to have information on how the solutions changed the organization."
"The company contact sent us the inventory information very late."
"They didn't send us all requested information."
"This challenge did not add or contribute to a deepening of the topics taught in this curricular unit."

Table 3. Resume of companies' responses to the inquiry.

Companies Main Appreciation about the Assignment
"The fact that the students have to analyze the investment and the return is very important."
"it's a pity we did not have more time to dedicate to you."
"Be persistent."
"Although the solutions presented may not be directly applicable, they are the source of new ideas."
"The challenge is a great way for future professionals to get to know our company."
"students must be more precise in what information they need us to provide."

Table 4. Resume of teachers' responses to the inquiry.

Teachers Main Appreciation about the Assignment
"The comprehension of the production processes and the application of life cycle assessment methodologies is better attained."
"The challenges also contributed to the promotion of student's transversal skills."
"In general, both students and companies consider that the challenges have an added value."
"the work helps to link classroom teaching with the factory shop floor practice."
"students begin by having a lot of resistance to this kind of assignment."
"To the majority of students, this was the first time that they had to produce and edit a video of this type."
"Three video deliveries ended being not so positive. A poster delivery could be proposed instead, contributing to a new skill request."
"Intermediate milestones were very important to promote a continues work in the assignment during the semester."
"arranging proper companies and products is complex."

4. Discussion

Globally speaking, students attribute value to the assignment and even suggest that more curricular unit assignments could be developed with interaction with the industrial sector. Although some initial difficulties understanding the overall assignment and some communication difficulties were reported by student and by companies, the overall sense is that the assignment created the conditions for students to acquire a critical and innovative sense in relation to the way that products are manufactured, being, therefore, better prepared to develop optimization projects, problem resolution in this technology area, and development of an entrepreneurial project that incorporates eco-design and eco-efficiency philosophies and may lead to the adoption of circular business models.

Most of the companies, if not all, pointed out the importance of merging the technical proposals with the required investment, as well as the training of staff and revenue of the investment. This was considered an added value of the assignment, and it was recommended that teachers reinforce this point in their classes along with the curricula. The openness of companies to receive students and interact with them after the first visit depended very much on their own availability, but company personnel always encouraged students to be pro-active and persistent to attain the information needed. Companies also recognized that the assignment was a good source of new ideas and solution proposals, besides it being a good opportunity for them to interact with potential future collaborators. These findings are well aligned with the literature [28,30].

The teachers and other staff involved with the application and management of the assignment consider it to be a good methodology to open students' mindsets to the challenges they will face in the near future in the workplace. This assignment promoted technical competencies related with product and production eco-design and ecoefficiency, but also competencies that were not yet developed within their academic path, such as a direct relationship with companies. Probably because of the originality of the assignment, the number of students reluctant to do the assignment was considerable. Nevertheless, at the end of the work, it was considered a positive experience by most. Some students still considered the assignment a waste of time and felt that it did not contribute to their academic development. The logistics of the assignment are complex. Not all companies are willing to receive a group of students and share with them their production details. Similar experiences have been reported elsewhere [38–42].

Intermediate milestones were very important to promote a working pace. The first three milestones were delivered in short length video format. To the majority of students, this was the first time that they had to produce and edit a video of this type. It was considered a good strategy to also promote new transversal skills. In the future, the third milestone may be requested in poster format instead of video, so skills in poster production can also be developed and the number of videos requested might be reduced. The mandatory blending of students from different courses and different backgrounds was not well accepted at first but was later considered a positive point.

In a nutshell, having in mind the questions formulated to evaluate the perception and appreciation of the learning and teaching outcomes by the involved agents, it can be considered that the assignment contributed to the proposed goals of addressing eco-design and eco-efficiency to promote a circular economy mindset and the integration of teaching and learning with the surrounding industry. Students were motivated by the assignment and were able to connect the subjects that they learned in the classroom to the challenge case and develop a better comprehension of the workplace, as well as improve their fundamental and transversal soft skills. Moreover, companies were satisfied by the technical results proposed and with their interactions with students. As new university business collaborations are successfully developed, their relationships will be improved.

5. Conclusions

Today's engineering and design students will be the builders of tomorrow's products. Sustainability is a responsibility of each inhabitant of planet Earth, but engineers and designers play an important role in this. It is, therefore, of vital importance that engineers develop competencies and a mindset to rethink products, decrease the resource use intensity, pursue ways to attain more efficient and cleaner production models, and convert linear business models to circular economy business models.

The proposed challenge contributed to the goal of integrating societal challenges within students' mindsets while developing their technical and transversal skills. The assignment created the condition for a real interaction between students and industrial agents, and through them, between the university and the companies, which have already led to new research collaborative projects. As reported by the literature, collaboration between universities and the different actors of the economic environment leads to a series of benefits that have a favorable impact on business competitiveness and university curricula [27–30,38–42].

It must be noticed that the findings presented here were acquired from one single application of the assignment model. It is intended to repeat the present model, with the incorporation of the improvements highlighted in this paper, in future academic years.

Funding: The research here presented was supported by the projects UID/EMS/00481/2019-FCT - FCT - Fundação para a Ciência e a Tecnologia; and CENTRO-01-0145-FEDER-022083 - Centro Portugal Regional Operational Program (Centro2020), under the PORTUGAL 2020 Partnership Agreement, through the European Regional Development Fund.

Acknowledgments: The author greatly thanks the openness of the companies involved.

Educ. Sci. **2019**, *9*, 126

Conflicts of Interest: The author declares no conflict of interest. The funders had no role in the design of the study; in the collection, analyses, or interpretation of data; in the writing of the manuscript, or in the decision to publish the results.

References

1. Bovea, M.D.; Pérez-Belis, V. A taxonomy of ecodesign tools for integrating environmental requirements into the product design process. *J. Clean. Prod.* **2012**, *20*, 61–71. [CrossRef]
2. Knight, P.; Jenkins, J.O. Adopting and applying eco-design techniques: A practitioners perspective. *J. Clean. Prod.* **2012**, *17*, 549–558. [CrossRef]
3. Pigosso, D.C.A.; Rozenfeld, H.; McAloone, T.C. Ecodesign maturity model: A management framework to support ecodesign implementation into manufacturing companies. *J. Clean. Prod.* **2013**, *59*, 160–173. [CrossRef]
4. Santoyo-Castelazo, E.; Azapagic, A. Sustainability assessment of energy systems: Integrating environmental, economic and social aspects. *J. Clean. Prod.* **2014**, *80*, 119–138. [CrossRef]
5. Roblek, V.; Meško, M.; Krapež, A. A Complex View of Industry 4.0. *SAGE Open* **2016**. [CrossRef]
6. Lopes de Sousa Jabbour, A.B.; Jabbour, C.J.C.; Godinho Filho, M.; Roubaud, D. Industry 4.0 and the circular economy: A proposed research agenda and original roadmap for sustainable operations. *Ann. Oper. Res.* **2018**, *270*, 273–286. [CrossRef]
7. Tseng, M.-L.; Tan, R.R.; Chiu, A.S.F.; Chien, C.-F.; Kuo, T.C. Circular economy meets industry 4.0: Can big data drive industrial symbiosis? *Resour. Conserv. Recycl.* **2018**, *131*, 146–147. [CrossRef]
8. Kalmykova, Y.; Sadagopan, M.; Rosado, L. Circular economy—From review of theories and practices to development of implementation tools. *Resour. Conserv. Recycl.* **2018**, *135*, 190–201. [CrossRef]
9. Korhonen, J.; Honkasalo, A.; Seppälä, J. Circular Economy: The Concept and its Limitations. *Ecol. Econ.* **2018**, *143*, 37–46. [CrossRef]
10. Korhonen, J.; Nuur, C.; Feldmann, A.; Birkie, S.K. Circular economy as an essentially contested concept. *J. Clean. Prod.* **2018**, *175*, 544–552. [CrossRef]
11. Huhtala, A. *Circular Economy: A Commentary from the Perspectives of the Natural and Social Sciences*; EASAC European Academies Science Advisory Council: Brussels, Belgium, 2015; pp. 1–18.
12. Despeisse, M.; Baumers, M.; Brown, P.; Charnley, F.; Ford, S.J.; Garmulewicz, A.; Knowles, S.; Minshall, T.H.W.; Mortara, L.; Reed-Tsochas, F.P.; et al. Unlocking value for a circular economy through 3D printing: A research agenda. *Technol. Forecast. Soc. Chang.* **2017**, *115*, 75–84. [CrossRef]
13. Prieto-Sandoval, V.; Jaca, C.; Ormazabal, M. Towards a consensus on the circular economy. *J. Clean. Prod.* **2018**, *179*, 605–615. [CrossRef]
14. Walker, S.; Coleman, N.; Hodgson, P.; Collins, N.; Brimacombe, L. Evaluating the Environmental Dimension of Material Efficiency Strategies Relating to the Circular Economy. *Sustainability* **2018**, *10*, 666. [CrossRef]
15. Corsi, I.; Fiorati, A.; Grassi, G.; Bartolozzi, I.; Daddi, T.; Melone, L.; Punta, C. Environmentally Sustainable and Ecosafe Polysaccharide-Based Materials for Water Nano-Treatment: An Eco-Design Study. *Materials* **2018**, *11*, 1228. [CrossRef] [PubMed]
16. Saidani, M.; Yannou, B.; Leroy, Y.; Cluzel, F.; Kendall, A. A taxonomy of circular economy indicators. *J. Clean. Prod.* **2019**, *207*, 542–559. [CrossRef]
17. Gallagher, J.; Basu, B.; Browne, M.; Kenna, A.; McCormack, S.; Pilla, F.; Styles, D. Adapting Stand-Alone Renewable Energy Technologies for the Circular Economy through Eco-Design and Recycling. *J. Ind. Ecol.* **2019**, *23*, 133–140. [CrossRef]
18. Figge, F.; Thorpe, A.S.; Givry, P.; Canning, L.; Franklin-Johnson, E. Longevity and Circularity as Indicators of Eco-Efficient Resource Use in the Circular Economy. *Ecol. Econ.* **2018**, *150*, 297–306. [CrossRef]
19. Navajas, A.; Uriarte, L.; Gandía, L.M. Application of Eco-Design and Life Cycle Assessment Standards for Environmental Impact Reduction of an Industrial Product. *Sustainability* **2017**, *9*, 1724. [CrossRef]
20. Whalen, K.A.; Berlin, C.; Ekberg, J.; Barletta, I.; Hammersberg, P. 'All they do is win': Lessons learned from use of a serious game for Circular Economy education. *Resour. Conserv. Recycl.* **2018**, *135*, 335–345. [CrossRef]
21. Sanganyado, E.; Nkomo, S. Incorporating Sustainability into Engineering and Chemical Education Using E-Learning. *Educ. Sci.* **2018**, *8*, 39. [CrossRef]

22. Coşkun, S.; Kayıkcı, Y.; Gençay, E. Adapting Engineering Education to Industry 4.0 Vision. *Technologies* **2019**, *7*, 10. [CrossRef]

23. Paravizo, E.; Chaim, O.C.; Braatz, D.; Muschard, B.; Rozenfeld, H. Exploring gamification to support manufacturing education on industry 4.0 as an enabler for innovation and sustainability. *Procedia Manuf.* **2018**, *21*, 438–445. [CrossRef]

24. Markopoulos, A.P.; Fragkou, A.; Kasidiaris, P.D.; Davim, J.P. Gamification in engineering education and professional training. *Int. J. Mech. Eng. Educ.* **2015**, *43*, 118–131. [CrossRef]

25. Müller, B.C.; Reise, C.; Seliger, G. Gamification in factory management education—A case study with Lego Mindstorms. *Procedia CIRP* **2015**, *26*, 121–126. [CrossRef]

26. Cebrián, G.; Junyent, M. Competencies in Education for Sustainable Development: Exploring the Student Teachers' Views. *Sustainability* **2015**, *7*, 2768–2786. [CrossRef]

27. Felder, R.M.; Brent, R.; Prince, M.J. Engineering Instructional Development: Programs, Best Practices, and Recommendations. *J. Eng. Educ.* **2011**, *100*, 89–122. [CrossRef]

28. Nottingham, P. The use of work-based learning pedagogical perspectives to inform flexible practice within higher education. *Teach. High. Educ.* **2016**, *21*, 790–806. [CrossRef]

29. Gabriel, B.F.C.C.; Valente, R.; Dias-de-Oliveira, J.A.; Neto, V.F.S.; De Andrade-Campos, A.G.D. A model for the effective engagement of all stakeholders in engineering education and its pilot implementation. *Eur. J. Eng. Educ.* **2018**, *43*, 950–966. [CrossRef]

30. Lester, S.; Costley, C. Work-based learning at higher education level: Value, practice and critique. *Stud. High. Educ.* **2010**, *35*, 561–575. [CrossRef]

31. International Organization for Standardization ISO 14040:2006. Available online: https://www.iso.org/standard/37456.html (accessed on 6 June 2019).

32. International Organization for Standardization ISO 14044:2006. Available online: https://www.iso.org/standard/38498.html (accessed on 6 June 2019).

33. Finkbeiner, M.; Inaba, A.; Tan, R.B.H.; Christiansen, K.; Klüppel, H.-J. The New International Standards for Life Cycle Assessment: ISO 14040 and ISO 14044. *Int. J. LCA* **2006**, *11*, 80–85. [CrossRef]

34. International Organization for Standardization ISO 14001:2015. Available online: https://www.iso.org/standard/60857.html (accessed on 6 June 2019).

35. International Organization for Standardization ISO 50001:2018. Available online: https://www.iso.org/standard/69426.html (accessed on 6 June 2019).

36. Sanders, A.; Elangeswaran, C.; Wulfsberg, J.P. Industry 4.0 implies lean manufacturing: Research activities in industry 4.0 function as enablers for lean manufacturing. *J. Ind. Eng. Manag.* **2016**, *9*, 811–833. [CrossRef]

37. Bai, C.; Satir, A.; Sarkis, J. Investing in lean manufacturing practices: An environmental and operational perspective. *Int. J. Prod. Res.* **2019**, *57*, 1037–1051. [CrossRef]

38. Şendoğdu, A.A.; Diken, A. A Research on the Problems Encountered in the Collaboration between University and Industry. *Procedia Soc. Behav. Sci.* **2013**, *99*, 966–975. [CrossRef]

39. Ankrah, S.; AL-Tabbaa, O. Universities–industry collaboration: A systematic review. *Scand. J. Manag.* **2015**, *31*, 387–408. [CrossRef]

40. Ivascu, L.; Cirjaliu, B.; Draghici, A. Business Model for the University-industry Collaboration in Open Innovation. *Procedia Econ. Financ.* **2016**, *39*, 674–678. [CrossRef]

41. Rajalo, S.; Vadi, M. University-industry innovation collaboration: Reconceptualization. *Technovation* **2017**. [CrossRef]

42. Huang, M.-H.; Chen, D.-Z. How can academic innovation performance in university–industry collaboration be improved? *Technol. Forecast. Soc. Chang.* **2017**, *123*, 210–215. [CrossRef]

© 2019 by the author. Licensee MDPI, Basel, Switzerland. This article is an open access article distributed under the terms and conditions of the Creative Commons Attribution (CC BY) license (http://creativecommons.org/licenses/by/4.0/).

education sciences

MDPI

Article

Enhancing Railway Engineering Student Engagement Using Interactive Technology Embedded with Infotainment

Sakdirat Kaewunruen[ID]

School of Engineering, The University of Birmingham, Birmingham B15 2TT, UK; s.kaewunruen@bham.ac.uk; Tel.: +44-121-414-2670

Received: 30 April 2019; Accepted: 13 June 2019; Published: 16 June 2019

check for updates

Abstract: Interactive learning technology is an emerging innovation for future communication-aided teaching and learning that could positively enhance students' engagement and intrinsic motivation. Due to the virtue of interactive communication, classrooms are now anticipated to enable a variety of interaction-based learning technologies with diverse infotainment (a subset of "serious play") integrated with practical enquiry-based projects and case studies for employability improvement. In this paper, a comprehensive review of various teaching and learning pedagogies is assessed. Their suitability and association with infotainment and interactive technology is discussed and highlighted. In addition, a recent research activity on interactive communication is presented to form a new teaching application using interactive technology and infotainment (or edutainment) appropriate for student engagement in railway geometry and alignment design classes. The development of the integrated interactive technology and infotainment was implemented and evaluated in a postgraduate railway engineering class. Questionnaires were used to survey students' experiences in the classes with and without the technology enhanced learning. The outcome clearly shows that students enjoyed and felt they were significantly engaged in the class with the new interactive resources. Their participation and learning performance increased. Despite the favourable outcomes, the flexibility and viability of using this interactive technology still largely depends on the students' background and their previous experience.

Keywords: student engagement; interactive technology; active learning; clickers; infotainment; edutainment; teaching approaches; railway engineering

1. Introduction

Many British academic institutions are nervous about the outcome of the National Student Survey (NSS). In recent years, many institutions have tried to increase their NSS score by enabling teaching excellence. The NSS outcomes have informed us that students live digitally and they love having digital and interactive technologies embedded in a classroom [1,2]. Some examples are videos (e.g., Panopto, YouTube), infographic social media (e.g., Twitter, Instagram), and other technologies (e.g., Canvas, Blackboard). The uses of interactive technologies have become even more important as more students adopt various technology devices in their lives. These trends prompt the necessity to develop and implement technology enhanced learning, which engage students both inside and outside the classroom. This development indeed responds directly to various elements of the United Kingdom Professional Standards Framework in the Digital Age, such as: A1: design and plan learning activity; A4: develop effective learning environments; K3: how students learn; K4: the use of appropriate learning technologies; and V2: promote participation and equality for learners [3]. The focus of this study is placed on a specialized course for higher education in the United Kingdom. The similarity of the concept can vary from one country or continent to another, depending on the culture, norm, and values embedded in their educational systems. However, such aspects are beyond the scope of this study.

A number of diverse information and communication technologies (ICT) have been adopted in higher engineering education sector. By enabling active learning, the common goals are to increase students' intrinsic motivation, promote participation, and enhance learning outcomes [4–7]. Despite the diversity of student backgrounds, levels of study, and class environments, many attempts have led to best practices for facilitating ICTs in combination with flexible and suitable teaching pedagogies in higher education. In order to improve the student learning experience, an array of various teaching pedagogies suitable for a class situation are often used [8–11]. The use of the ICTs can, thus, benefit the adaptive teaching strategies and engage all students with different backgrounds.

Previous case studies in higher education institutions have demonstrated many benefits of interactive technology, such as clickers, student response systems via smart phones or tablets, interactive board, etc. [12–14]. The studies based on only student perceptions (opinion-based survey) found that students enjoy using the interactive technology and become much more engaged. With respect to clickers, Beatty et al. [15], Martyn [16], and Bruff [17] also found that clickers provide an anonymity for students to actively participate in class without fear of public humiliation; the tool creates "fun" and "game-based learning" that engages students more than traditional teaching. The peer game-based instruction could support deep learning [18]. This could be done by developing a class quiz for students to work individually or as a team to compete for the most appropriate, most-efficient, lowest-cost, or lowest-carbon-footprint solutions. Based on a number of similar findings in various cases [12,19,20], the integration of interactive technology and infotainment (a teaching medium both to entertain and to inform) can be complementarily established. The use of entertainment for an interactive teaching pedagogy has been acknowledged worldwide. For example, Jenkin [21] fully supported "fun" in teaching from his statement "play not only aids children's mental and physical health, it teaches them risk taking and problem solving skills, promoting imagination, independence, and creativity." This nature could be applied to higher education learners. In addition, clickers could be used as an alternative tactic to traditional class discussion in order to enhance active learning pedagogy [16,22,23].

Current teaching approaches within higher education uses adaptive blended learning, where students can attain a combination of teaching methods or pedagogies [24]. These adaptive approaches imply common practices in tailored teaching and learning, and have been supported by researches [25–28]. They are recommended to best engage students and to promote personalised technology enhanced learning. In fact, these approaches resonate with the active learning pedagogy, where students are actively engaged to learn through discovering, processing, and applying knowledge [29,30]. The forms of active learning can be cooperative learning or small group teaching [31,32], peer and group discussions [20], simulations and games [33], student response assistance, clickers [16,25], and so on. Despite the pros and cons of each form [34–37], there is an agreement [38–40] that the growth of personal technology integration in classrooms has been paramount and inevitable in higher education. In addition, many educators [19,41–46] have vigorously advocated that active learning through personalised responsive technology is one of the best teaching pedagogies and one of the most suitable practices for digital-native engineering students, especially in the 21st century. Many have also adopted it for a wide range of classes and leaners [47].

Clickers are personal response technology enhanced learning. They actively engage students during the entire class period. The prompt responses from clickers can gauge the level of understanding of the class so that lecturers can provide prompt feedback and enable adaptive teaching styles and blended pedagogies, resulting in a suitable harmonised student-centred learning environment. Martyn [16] supported this technology, stating that "clickers allow students to engage and provide input without fear of public humiliation and without having to worry about more vocal students dominating the discussion". However, she argued whether the clickers could actually help improve student learning outcome. Her study based on small class groups (about 20 students each) showed that clickers did not improve the learning outcomes when compared with another active learning pedagogy (using traditional class discussion). In contrast, recent findings based on larger sizes of class (>50 students) suggest otherwise [11,48,49]. They found that clickers consistently produced better learning outcomes as they play a significant role in cognitive learning; and they promote fact retention, even though they may impede conceptual understanding. Also, clickers seem to positively influence the intrinsic motivation and improve grades or final marks of students. These recent findings confirmed that one size does not fit all. Active learning tactics, such as group discussion or collaborative learning, might not be effective for larger class sizes [24]. Thus, personal responsive technology, such as clickers or smart phones, has been a revolutionary and inevitable teaching assistant tool in higher education, especially when looking towards the future. In the present day, smart phones are affordable and have been used in various classes.

Despite the popularity of clickers in university teaching, a critical literature review shows that the use of audience response systems (or clickers) is still limited in engineering fields [9,10,50,51] and is non-existent in railway engineering education. In this study, the learning activity and resources integrated with clickers have been developed. This unprecedented study explores the possible relationships between the interactive technology integrated with infotainment and student engagement in postgraduate railway engineering education. The educational development has been targeted at postgraduate students with different professional and demographic backgrounds. The lecture topic of track geometry and alignment design with over 50 students has been chosen particularly since many previous students had considered it as one of the most difficult topics in railway infrastructure modules. Many students were lacking confidence in applying the knowledge, demotivated, and disengaged from the learning process. The goal of this development is to understand if the clickers can better engage students in the seminar-style lecture, resulting in the enhancement of students' intrinsic motivation through active learning pedagogy. This study highlights the comparison of the student perception-based surveys considering the result in 2016 (using traditional group and class discussions) and its counterpart in 2017 (using clickers and infotainment). The work is aimed at improving the quality of higher education with respect to a railway engineering program by using an interactive technology and "fun" educational content to improve students' engagement, despite a common thought that engineering is a serious profession. The student learning performance based on an oral assessment of a group of final-year students from both cohorts is also evaluated to understand the role of interactive technology on the student learning outcome in a postgraduate railway infrastructure engineering module.

2. Development of Learning Activity and Resources with Integrated Learning Technology

The railway infrastructure module is a core module for the postgraduate programs MSc Railway Systems Engineering and Integration and MSc Railway Risk and Safety Management at the University of Birmingham. About 50 students or more enroll in this module every year, forming large-size classes for a masters level module. This module consists of a variety of lecture topics delivered by internal and external guest lecturers. The use of Canvas (digital communication and online learning materials) and Panopto (video records) have already been exploited to foster the digital learning process for students, starting in 2015. However, a constraint that discourages lecturers from utilising Panopto is that proprietary videos and multi-media are often used in class to enhance experiential-based learning pedagogy. A number of videos, simulations, and multi-media belong to the rail industry and private companies. For educational purposes, all classes are, however, recorded via Panopto to aid student learning experience. Reportedly, many students very much enjoy and appreciate using Panopto [1]. For this study, a specific topic has been chosen using the author's teaching responsibility in order to quantify the students' responses and learning outcome. With respect to the topic of track geometry and alignment design, the author has been solely responsible for the class lectures since 2015. Many past and previous students had negative experiences regarding the topic as it has been one of the most difficult lectures. They stated that they were not confident enough to apply the knowledge, were demotivated, and disengaged. A large number of students have failed the examination in the past. These issues have inspired the lecturer to adopt an active learning pedagogy to enhance the student experience. In 2016, industry case studies (project- and enquiry-based learning) were used in class for group work and class discussions (see Appendix A), aiming to enhance employability skills, as shown in Table 1. The seminar-style lecture was then developed to enact active learning integrated with the project-based learning approach [52] and multi-media to enhance the experiential learning of the students [53,54].

Table 1. Engineering employability skills required by employers, adopted from Kaewunruen [55].

U.K. [56]	Singapore [57]	Japan [57]
- New and specific technical skills - Computer literacy and IT skills - Multi-skilling and greater -flexibility - The ability to deal with change - An ability to continue learning, re-skilling - Communication skill - Team working and getting on with others, including being able to work in self-managed teams - problem-solving and diagnosis - "Whole system" thinking - Organisation and management	- Workplace literacy and numeracy - IT and Technology - Problem solving - Initiative and enterprise - Communication and Relationship - Lifelong learning - Globalisation - Self-management - Workplace-related life skills - Health and workplace safety	- Communication skills - Problem solving - Goal-setting skill - Personal presentation skills - Visioning skills - IT and computer - Leadership - Self-assessment skills

Interactive technology (using TurningPoint Technology's clickers) embedded with infotainment (game-based questions and fun quizzes) in the class slides (using MS Powerpoint) have, thus, been further developed as a new learning resource since late 2016 (see Appendix B). The development has fully recognised that students have different professional and cultural backgrounds and conceptions of learning; students' expectations and conceptions of learning shape how they approach study tasks and class activities. Several multi-media have been used in class to offer story telling entertainment to enhance the learning experience and create deeper learning. Clickers associated with infotainment-based questions have been developed and used to engage all students (from passive to active learning) and obtain class feedback for adaptive teaching tactics. The use of clickers has been designed to provide a break between key technical concepts to gauge students' understanding of the materials, to engage every student via game approach, as well as short class discussions, and to revive their energy and attention by infotainment. A number of multiple choice questions were set to obtain student responses. The questions were designed to allow students to have peer or group discussions before or after the vote. These discussions are critically important as they enhance the communication skills of individual students. Data received from students were acquired by a dongle plugged into a PC in front of the class. These data were then saved onto the powerpoint for discussion and analyses in this study.

3. Evaluation Plan

The effectiveness of interactive teaching and student engagement are evaluated using perception-based questionnaires (see Appendix C). All students in class were given the questionnaires immediately after each lecture to evaluate their fresh perspectives on the teaching styles and their feelings towards the level of engagement. The questionnaire is aimed at echoing the students' feedback about the teaching pedagogy (active learning), the use of digital technology, the advantages and preference of using technology (i.e., Panopto), the preference towards teaching styles, delivery mode, quality of teaching materials, and the use of online communication (i.e., Canvas). These perception-based questions were later correlated with the demographic data of students, including gender, English proficiency, cultural background, professional experience, and digital-native orientation. Almost every student (approximately 40–50 students) each year took part in the surveys.

Student learning outcomes are also measured by taking the score from the oral examination of students (based on the curriculum of MSc Railway Risk and Safety Management) at the end of the year 2016/2017 (taken in May 2017) using situation-based enquiries. Out of 11 candidates (the whole cohort), five students (part timers) took the class in 2016 (traditional class discussion) and the rest were in 2017 lecture (interactive technology). This was the only formal evaluation available to critically evaluate "student learning", as all exams are normally conducted at the end of the program (e.g., end of academic year). The students were given the case scenario an hour prior to the oral exam. In the exam, the students had 10 min to read exam questions and prepare answers. Then, they were interviewed using the given questions and evaluated from the completeness of answers and application of knowledge to the case scenario. Two experienced assessors marked each student independently to assure fairness and comprehensiveness. The whole interview examination takes 30 min for each candidate.

4. Learning Design Flow Map

The new lecture developed in 2017 has embedded interactive technology and infotainment to activate students' active learning and deeper thinking. The use of clickers is found to be very appropriate for this lecture (>50 students), since Hoekstra and Mollborn [58] also found that seminar-style discussions or lectures become more difficult when the class contains more than 35 students. Based on a teaching conceptual analysis in accordance with Blasco-Arcas et al. [7], Figure 1 shows the learning concept framework of the interactive lecture.

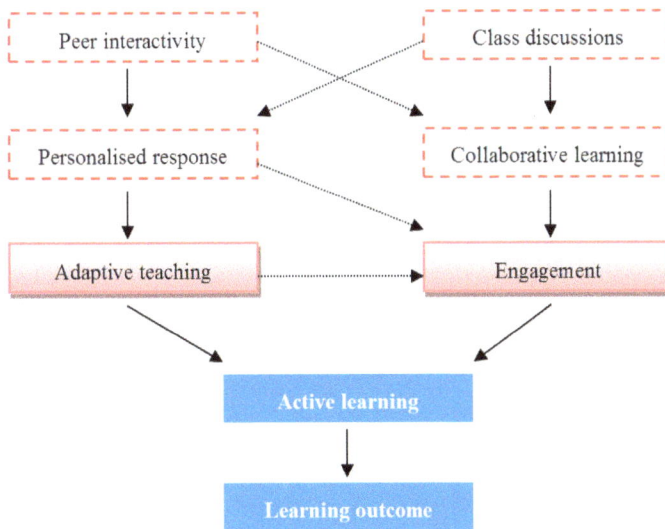

Figure 1. Learning concept framework using interactive technology with infotainment.

The analysis shows the flow of teaching tactics and pedagogies to enhance active learning and student learning outcome [59]. The use of clickers increases interactivity, adaptive teaching, and then engagement, stimulating students to participate in active learning for better learning outcomes. This pedagogical rationale underpins the implementation of interactive technology with infotainment in class.

The learning sequence in the railway engineering module can be designed as illustrated in Figure 2 [60–63]. As a result of a successful class in 2016, the development of learning resources to integrate interactive technology and infotainment was to foster individual responses and engagement in all students (in the United Kingdom, European Union, and internationally). The novel feature of clickers associated with infotainment questions was incorporated in various learning activities. The initial step was to identify audiences and to tailor teaching tactics and styles. For example, more than 70% of students were not from civil engineering fields, so further explanations of terminologies and concepts were a must. To add authenticity and engagement, small group discussions and feedback were also enacted together with individual responses and "fun-oriented" questions. The learning design has proven to be flexible and adaptive in that blended teaching tactics have been used in the class. The teaching materials and Panopto video recordings are also available on Canvas for students to review for exams at the end of the year.

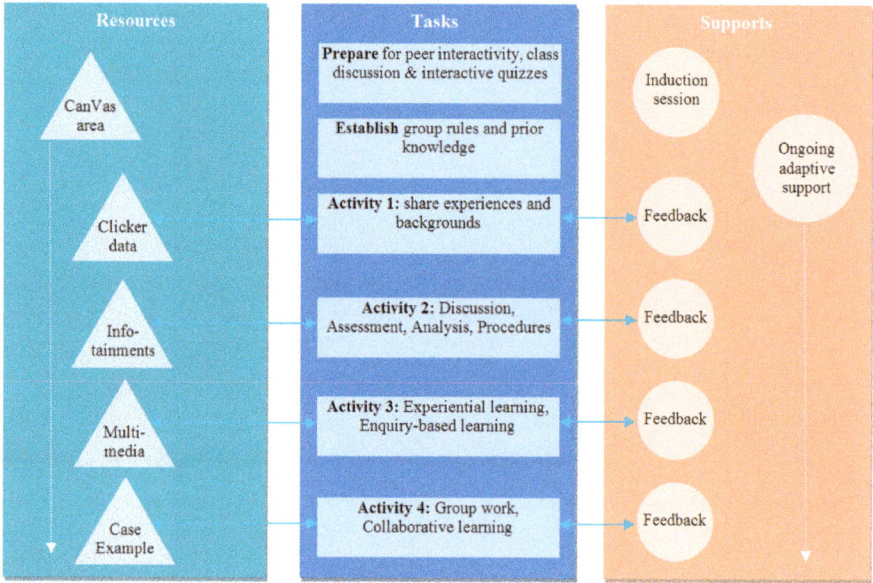

Figure 2. Learning design sequence using interactive technology with infotainment, adopted from McLinden and Hinton [60].

5. Results and Discussion

Participation data of both academic years 2016 and 2017 were gathered from the questionnaires. These data were used to determine student demography and perception. Attendance in 2017 was slightly higher than that in 2016. Note that student attendance does count toward the final mark. Table 2 shows the demographic comparison of student cohorts in both years. This confirms that rail is a male-dominated industry and the majority of students are international with some professional experience. The majority of the students (around 70%) are of mature age (between 25 to 35 years old), which also implies that they are more responsible for their own learning and target career paths. As a result, class discussions should be made relevant and sensitive to various groups of students in order to draw on students' previous experiences and exposure for class discussions and interactions. The results in Table 2 show an interesting outcome that both 2016 and 2017 cohorts did not enjoy using online learning technology. They prefer face to face learning and interaction, which also results in human connection and networking.

Table 2. Demographics of student backgrounds (all combined cohorts who took the class).

Demography	2016 Cohort * (Traditional Class Discussion)	2017 Cohort * (Interactive Technology)
Male	60%	78%
Female	40%	22%
Native English speaker	38%	45%
Have professional experience	56%	58%
Enjoy using technology	38%	28%
Total number of survey response	35	46
Total number of enrolment	42	51

* Note: these numbers were based on the actual students in the class. These students may or may not take the end-of-program exams for graduation as they may prolong their studies.

Note that the survey was planned using the university standard practice. The questionnaires were developed in consultation with the university's Higher Education Office to assure that the specific research objectives can be determined for educational improvement without jeopardizing students' data privacy. Based on the survey results, student perceptions of using class discussion and clickers are shown in Figure 3. The student surveys in 2016 and 2017 show that students enjoyed the interactive teaching class slightly more and felt they were significantly engaged in the class with the new interactive resource. The average scores are consistently higher for the 2017 student cohort who had used interactive technology (compared with those in 2016). This implies that a positive experience could be attained when using the interactive technologies. Considering the adaptive teaching approach, the students felt that the subject is less difficult and they needed lesser online support (such as simulation or video record). This insight can be derived from the apparent change in student perceptions. The survey data also implies that students perceive considerable value in interactive technology with infotainment, especially when the students can interact with smart phone devices. The score on the preference of using smart phones is higher than interacting via Canvas (an online class communication tool).

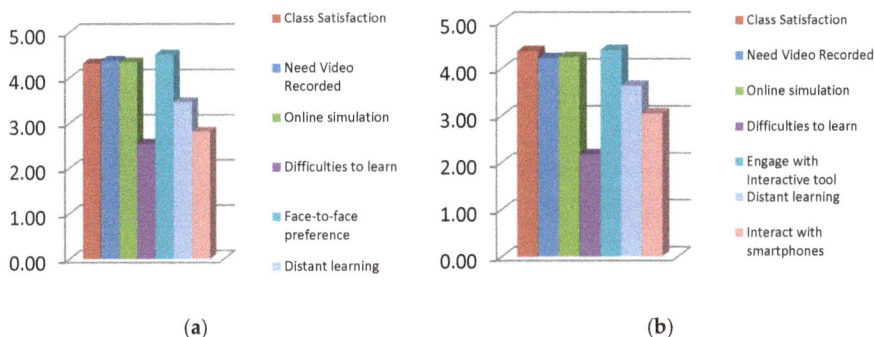

(a)   (b)

Figure 3. Student perceptions (**a**) average values in 2016 and (**b**) average values in 2017. Note that the mean standard deviation is about ±1.07 in 2016 and about ±0.92 in 2017.

Despite the favorable outcomes, Table 3 displays some student comments from the 2017 cohort. The mixed comments from positive and negative responses can be observed. Based on the free comment feedback obtained from the survey, the adaptability, flexibility, and viability of using this interactive technology still largely depends on the students' backgrounds and their previous experience. These comments reflect the need to identify the audiences early on and to tune the teaching tactics to accommodate the learning styles preferred by different groups of students. For example, native English speaking students would enjoy more class discussions in combination with the interactive technology, since vocal discussions can be considered their strength. However, very positive comments were obtained frequently from students of various backgrounds. Based on students' verbal and written feedback, non-native English students tend to significantly complement the use of clickers for engagement and entertainment, since public speaking and discussion is one of their weaknesses or sources of discomfort. The lesson learnt from the use of interactive technology (or clickers) was that a teacher cannot overestimate the potential of technology without being responsive to the students and the cohort. Again they say, "one size does not fit all". The teaching process must be ready to be adaptively and optimally tailored to suit the combination of the class and each student's need.

Table 3. Selection of student comments from the 2017 cohort using clickers.

Student Comments
● Investigate how people could be encouraged to interact verbally.
● Great effort, well done.
● Remove names from clicker questions. Doing this actually create 2 questions and is confusing. Need better definitions of alignment and geometry—very confusing.
● I would have liked to work through the work example with the class, i.e., for the lecturer to show us afterward.
● Ensure video lecture is available online.
● I liked the interactive tool. I am a designer and it was good to hear it from an academic view.
● Needs video to give real view of the materials.
● I prefer if concepts were detailed better in lectures.
● Use of technology was good. Kept me engaged.

At the end of the year 2016/2017, student learning outcomes could be measured by considering the score on the end-of-year oral examination of students (based on the curriculum of MSc Railway Risk and Safety Management). All students in the program must undergo the oral examination (which could permit the lecturer to assess the round understanding of the topic issue). This study has adopted the oral examination for the baseline, as other students in other program will be required to take the final exam combining various parts or sub-modules (e.g., train-track interaction, transit space, etc.) within the infrastructure exam. As such, the final exam results could not be used to distinguish the effectiveness of the teaching tactic in this paper. Therefore, only the oral examination has been chosen for further analysis. The oral examination was carried out by two faculty members by asking technical questions and listening to the student responses. A score for each student was awarded by each faculty member and then an average mark was used for the final outcome. Figure 4 illustrates the comparison of overall academic merit ranking. Note that A implies "excellent", with a mark over 70%; B is "good", with a mark between 60% and 70%; C is "fair", with a mark between 50% and 60%; D is "poor", between 40% and 50%; and F is a fail grade (under40%). As discussed earlier in Section 4, Figures 1 and 2 show the educational design of the classes in both years (2016 and 2017). The outcome in this study (as in Figure 4) clearly distinguishes the learning outcome of students from different teaching techniques.

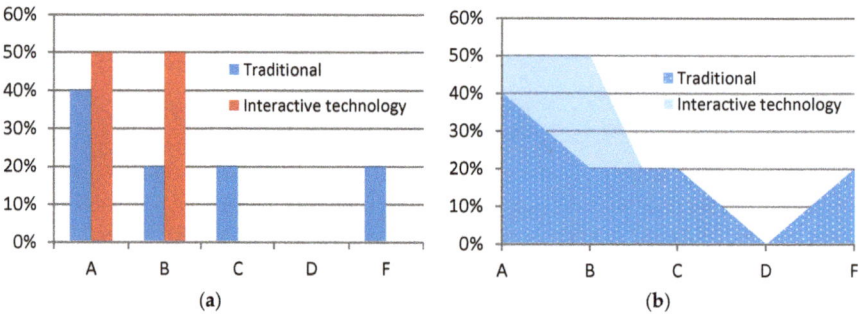

Figure 4. Comparison of student learning outcomes: (a) comparative bar chart of academic merit; and (b) area of influence on overall academic merit.

Based on the number of candidates (11 May 2017), Figure 4 shows that students who used interactive technology tend to outperform those students using traditional teaching tools. All students (100%) who used clickers could tackle the oral examination with a mark well above 60%, whilst only 60% of the students who used the traditional class discussion approach could achieve a mark above 60% (B level). It is apparent that about 40% of students who did not use the interactive teaching approach performed poorly and up to 20% of them could also fail the oral exam. Based on this exam result, it shows that the interactive technology has a positive impact on the overall student attitudes and student engagement, resulting in higher learning performance. The next generation of students will rely on digital classrooms and the interactive teaching must be adapted to function with online technology in the future. Note that this study is based on the assumption that students do not suffer from poor English proficiency or any cultural barriers during the classes, survey examination, and other assessments. The cultural aspects are beyond the scope of this study.

6. Conclusions

An interactive technology (clickers) was used in conjunction with infotainment in a postgraduate railway engineering module in 2017. This development responds to many elements of the United Kingdom Professional Standards Framework in the Digital Age. Due to the virtue of interactive communication, the classroom is now expected to enable a combination of interaction-based learning technologies integrated with practical enquiry-based projects and case studies for improving employability skills. The application of clickers is aimed at improving student engagement, which could lead to intrinsic motivation and active learning, and to eventually improve the learning outcomes of the students. This paper shows that the use of clickers can prompt the lecturer to tailor the teaching styles and tactics to improve the understanding of the students. From this study, a number of conclusions can be identified:

- Blended infotainment and interactive technology can form an adaptive teaching pedagogy to increase student engagement in the railway geometry and alignment design class.
- The student surveys highlight the enjoyment, stimulation, and engagement of students in class, leading to improved learning performance.
- Students also perceive added value of interactive technology integrated with infotainment, renewing their learning participation.
- Despite the positive outcomes, the flexibility and viability of using this interactive technology still largely depends on the nature of the audiences.

It should be noted that this study was based on a limited number of sample data. It was also the first time the author had facilitated the class using clickers. On this ground, future work will include a more adaptive teaching style to enact active learning for all students and to further analyse the final exam marks of the students. In addition, another type of interactive technology, such as augmented reality, will be considered to improve the student engagement in class.

Author Contributions: Conceptualization, S.K.; methodology, S.K.; software, S.K.; validation, S.K.; formal analysis, S.K.; investigation, S.K.; resources, S.K.; data curation, S.K.; writing—original draft preparation, S.K.; writing—review and editing, S.K.; visualization, S.K.; project administration, S.K.; funding acquisition, S.K.

Funding: This project has received funding from the European Union's Horizon 2020 research and innovation programme under the Marie Skłodowska-Curie grant agreement No 691135 "RISEN: Rail Infrastructure Systems Engineering Network". The APC was funded by The University of Birmingham Library's Open Access Fund.

Acknowledgments: S.K. wishes to thank the Australian Academy of Science and the Japan Society for the Promotion of Sciences for his Invitation Research Fellowship (Long-term) at the Railway Technical Research Institute and The University of Tokyo, Japan. The authors wish to gratefully acknowledge the financial support from the European Commission for H2020-MSCA-RISE Project No. 691135 "RISEN: Rail Infrastructure Systems Engineering Network", which enables a global research network that tackles the grand challenges [64] in railway infrastructure resilience and advanced sensing in extreme events (www.risen2rail.eu). The technical review and constructive comments by Danielle Hinton and Marios Hadjianastasis are gratefully acknowledged.

Conflicts of Interest: The authors declare no conflict of interest.

Appendix A

Sample of teaching slides

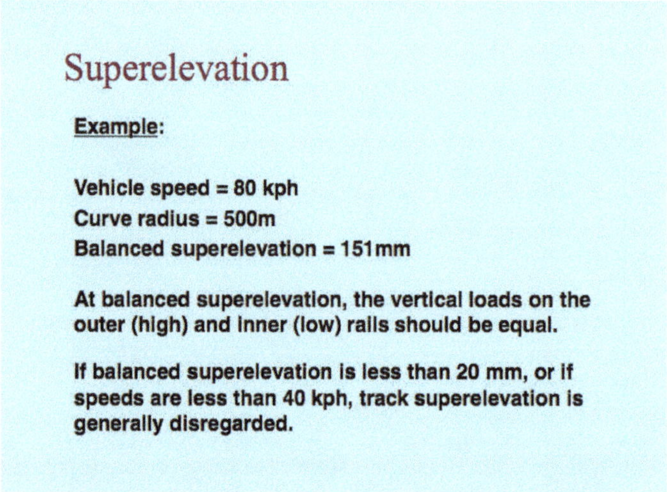

Figure A1. Lecture Slide in 2016 (Traditional Class Discussion).

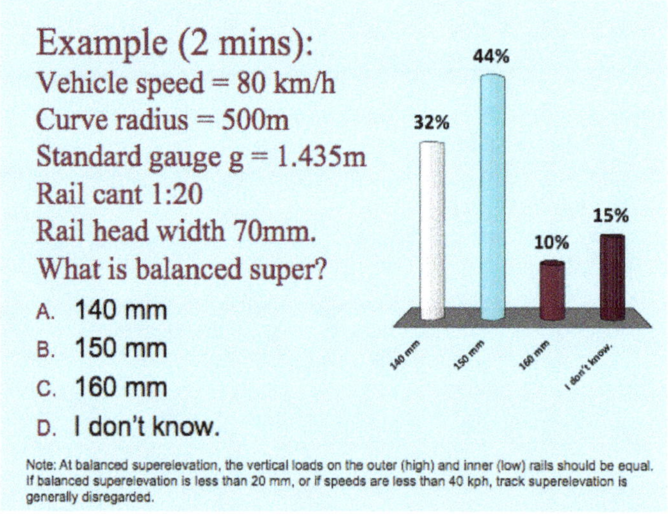

Figure A2. Interactive Lecture Slide in 2017 (Interactive Technology with Infotainment).

Appendix B

Case Study for Class Discussion

Infrastructure Module

Rail Geometry and Alignment Design

Group Discussion – Work in Group of 3 students

(10 mins = 8 mins discussion + 2 mins reporting)

A train derailed at a yard in Manchester. The derailment occurred in a hot summer while the empty freight train run at the speed of 20km/h. The yard manager investigated the scene and reported that a freight train wheel climbed over a shallow transition (500m radius) of the open plain track, commencing from the high rail (outer rail). The yard manager complained that 'this derailment was because of *wrong* track geometry design'. This rumour was somehow leaked to a daily morning herald.

You are a stellar Birmingham alumnus and have recently been appointed as the CEO of the railway company. A news reporter from Chanel **XXX** is live on a teleconference interviewing you about this incident.

The reporter asked you "Can you confirm that this derailment was due to wrong curve design?"

What would be your reply?

Yes or No?

Based on your previous study back in your stellar year in Birmingham, you know that the risk of wheel climbing derailment depends on critical lateral and vertical force ratio (L/V):

$$\frac{L}{V} = \frac{\tan\delta - \mu}{1 + \mu\tan\delta}$$

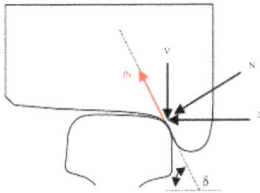

Figure A3. Case discussion on the potential causes of a train derailment

Appendix C

A sample of questionnaires to understand student perception (2017)

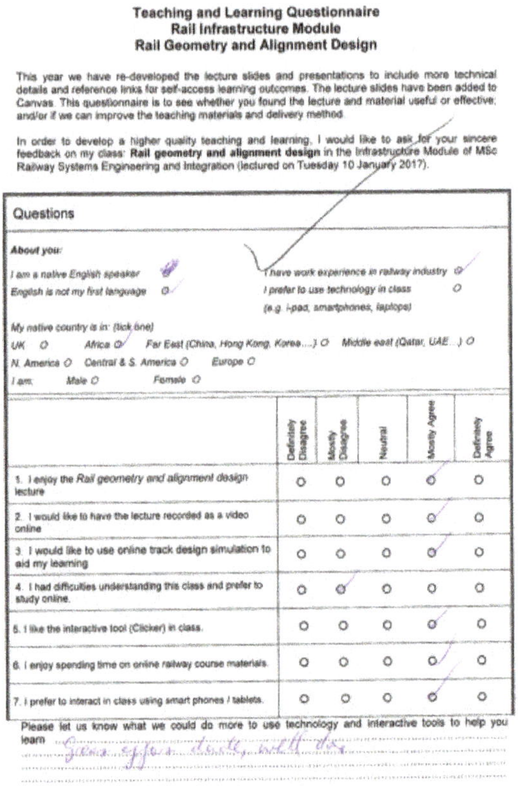

Teaching and Learning Questionnaire
Rail Infrastructure Module
Rail Geometry and Alignment Design

This year we have re-developed the lecture slides and presentations to include more technical details and reference links for self-access learning outcomes. The lecture slides have been added to Canvas. This questionnaire is to see whether you found the lecture and material useful or effective; and/or if we can improve the teaching materials and delivery method.

In order to develop a higher quality teaching and learning, I would like to ask for your sincere feedback on my class: **Rail geometry and alignment design** in the Infrastructure Module of MSc Railway Systems Engineering and Integration (lectured on Tuesday 10 January 2017).

Questions					
About you:					
I am a native English speaker ☑ I have work experience in railway industry ○					
English is not my first language ○ I prefer to use technology in class ○ (e.g. i-pad, smartphones, laptops)					
My native country is in: (tick one)					
UK ○ Africa ○ Far East (China, Hong Kong, Korea ....) ○ Middle east (Qatar, UAE ...) ○					
N. America ○ Central & S. America ○ Europe ○					
I am: Male ○ Female ○					

	Definitely Disagree	Mostly Disagree	Neutral	Mostly Agree	Definitely Agree
1. I enjoy the *Rail geometry and alignment design* lecture	○	○	○	⊙	○
2. I would like to have the lecture recorded as a video online	○	○	○	⊙	○
3. I would like to use online track design simulation to aid my learning	○	○	○	⊙	○
4. I had difficulties understanding this class and prefer to study online.	○	⊙	○	○	○
5. I like the interactive tool (Clicker) in class.	○	○	○	⊙	○
6. I enjoy spending time on online railway course materials.	○	○	○	⊙	○
7. I prefer to interact in class using smart phones / tablets.	○	○	○	⊙	○

Please let us know what we could do more to use technology and interactive tools to help you learn ....

Figure A4. Example of student feedback

References

1. Armour, K. *Message to All Staff from Pro Vice Chancellor (Education), E-mail Communication to All Staff on 13 September 2016*; The University of Birmingham: Birmingham, UK, 2016.
2. Kaewunruen, S. Underpinning systems thinking in railway engineering education. *Australas. J. Eng. Educ.* **2018**. [CrossRef]
3. UKPSF (UK Professional Standards Framework). *The UK Professional Standards Framework for Teaching and Supporting Learning in Higher Education*; The Higher Education Academy, Guild HE, Universities UK: London, UK, 2011.
4. Guthrie, R.; Carlin, A. Waking the Dead: Using interactive technology to engage passive listeners in the classroom. In Proceedings of the Tenth Americas Conference on Information Systems, New York, NY, USA, 6–8 August 2004; pp. 2952–2959.
5. Ribeiro, L.R.C.; Mizukami, M.D.G.N. Problem-based learning: A student evaluation of an implementation in postgraduate engineering education. *Eur. J. Eng. Educ.* **2005**, *30*, 137–149. [CrossRef]
6. Kirkwood, A. E-learning: You don't always get what you hope for. *Technol. Pedagog. Educ.* **2009**, *18*, 107–121. [CrossRef]
7. Blasco-Arcas, L.; Buil, I.; Hernández-Ortega, B.; Javier Sese, F. Using clickers in class. The role of interactivity, active collaborative learning and engagement in learning performance. *Comput. Educ.* **2013**, *62*, 102–110. [CrossRef]
8. Caldwell, J.E. Clickers in the Large Classroom: Current Research and Best-Practice Tips. *CBE Life Sci. Educ.* **2007**, *6*, 9–20. [CrossRef] [PubMed]
9. Schmidt, B. Teaching engineering dynamics by use of peer instruction supported by an audience response system. *Eur. J. Eng. Educ.* **2011**, *36*, 413–423. [CrossRef]

10. Han, J.H.; Finkelstein, A. Understanding the effects of professors' pedagogical development with Clicker Assessment and Feedback technologies and the impact on students' engagement and learning in higher education. *Comput. Educ.* **2013**, *65*, 64–76. [CrossRef]
11. Brady, M.; Seli, H.; Rosenthal, J. "Clickers" and metacognition: A quasi-experimental comparative study about metacognitive self-regulation and use of electronic feedback devices. *Comput. Educ.* **2013**, *65*, 56–63. [CrossRef]
12. Crossgrove, K.; Curran, K.L. Using Clickers in Nonmajors- and Majors-Level Biology Courses: Student Opinion, Learning, and Long-Term Retention of Course Material. *CBE Life Sci. Educ.* **2008**, *7*, 146–154. [CrossRef]
13. Kennewell, S.; Tanner, H.; Jones, S.; Beauchamp, G. Analysing the use of interactive technology to implement interactive teaching. *J. Comput. Assist. Learn.* **2007**, *24*, 61–73. [CrossRef]
14. Bojinova, E.; Oigara, J. Teaching and learning with clickers in higher education. *Int. J. Teach. Learn. High. Educ.* **2013**, *25*, 154–165.
15. Beatty, I.D.; Gerace, W.J.; Leonard, W.J.; Dufresne, R.J. Designing effective questions for classroom response teaching. *Am. J. Phys.* **2006**, *74*, 31–39. [CrossRef]
16. Martyn, M. Clickers in the classroom: An active learning approach. In Proceedings of the Annual Conference EDUCAUSE, Philadelphia, PA, USA, 31 October–3 November 2007.
17. Bruff, D. *Teaching with Classroom Response Systems: Creating Active Learning Environments*; Jossey-Bass: San Francisco, CA, USA, 2009.
18. Mazur, E. *Peer Instruction: A User's Manual*; Pearson: New York, NY, USA, 1997.
19. McKnight, K.; O'Malley, K.; Ruzic, R.; Horsley, M.K.; Franey, J.J.; Bassett, K. Teaching in a digital age: How educators use technology to improve student learning. *J. Res. Technol. Educ.* **2016**, *48*, 194–211. [CrossRef]
20. Johnson, D.W.; Johnson, R.T.; Smith, K.A. *Active Learning: Cooperation in the College Classroom*; Interaction Book Company: Edina, MN, USA, 1998.
21. Jenkin, M. Learning through Play: Pedagogy, Challenges and Ideas—Live Chat, The Guardian. 2013. Available online: https://www.theguardian.com/teacher-network/teacher-blog/2013/feb/15/learning-play-imaginative-inquiry-teaching-schools-live-chat (accessed on 1 January 2019).
22. Laxman, K. A study on the adoption of clickers in higher education. *Australas. J. Educ. Technol.* **2011**, *27*, 1291–1303. [CrossRef]
23. O'Flaherty, J.; Phillips, C. The use of flipped classrooms in higher education: A scoping review. *Internet High. Educ.* **2015**, *25*, 85–95. [CrossRef]
24. Bernard, R.M.; Borokhovskil, E.; Schmid, R.F.; Tamim, R.M.; Abrami, P.C. A meta-analysis of blended learning and technology use in higher education: From the general to the applied. *J. Comput. High. Educ.* **2014**, *26*, 87–122. [CrossRef]
25. Elliott, C. Using a Personal Response System in Economics Teaching. *Int. Rev. Econ. Educ.* **2003**, *1*, 80–86. [CrossRef]
26. Hannay, M.; Fretwell, C. The higher education workplace: Meeting the needs of multiple generations. *Res. High. Educ. J.* **2011**, *10*, 1–12.
27. Junco, R.; Heibergert, G.; Loken, E. The effect of Twitter on college student engagement and grades. *J. Comput. Assist. Learn.* **2011**, *27*, 119–132. [CrossRef]
28. Ferreri, S.P.; O'Connor, S.K. Redesign of a large lecture course into a small-group learning course. *Am. J. Pharm. Educ.* **2013**, *77*, 13. [CrossRef]
29. McKinney, K.; Heyl, B. (Eds.) *Sociology through Active Learning*; SAGE/Pine Forge Press: Thousand Oaks, CA, USA, 2008.
30. Kassens-Noor, E. Twitter as a teaching practice to enhance active and informal learning in higher education: The case of sustainable tweets. *Act. Learn. High. Educ.* **2012**, *13*, 9–21. [CrossRef]
31. Machemer, P.L.; Crawford, P. Student perceptions of active learning in a large cross-disciplinary classroom. *Act. Learn. High. Educ.* **2007**, *8*, 9–30. [CrossRef]
32. Cavanagh, M. Students' experiences of active engagement through cooperative learning activities in lectures. *Act. Learn. High. Educ.* **2011**, *12*, 23–33. [CrossRef]
33. Lean, J.; Moizer, J.; Towler, M.; Abbey, C. Simulations and games: Use and barriers in higher education. *Act. Learn. High. Educ.* **2006**, *7*, 227–242. [CrossRef]

34. Bingimals, K.A. Barriers to the successful integration of ICT in teaching and learning environ-ments: A review of the literature. *Eurasia J. Math. Sci. Technol. Educ.* **2009**, *5*, 235–245. [CrossRef]
35. Davis, M. Barriers to Reflective Practice: The Changing Nature of Higher Education. *Act. Learn. High. Educ.* **2003**, *4*, 243–255. [CrossRef]
36. Walker, S.E. Active learning strategies to promote critical thinking. *J. Athl. Train.* **2003**, *38*, 263–267.
37. Yazedjian, A.; Boyle Kolkhorst, B. Implementing Small-Group Activities in Large Lecture Classes. *Coll. Teach.* **2007**, *55*, 164–169. [CrossRef]
38. Chan, T.F.I.; Borja, M.; Welch, B.; Batiuk, M.E. Predicting the probability for faculty adopting an audience response system in higher education. *J. Inf. Technol. Educ. Res.* **2016**, *15*, 395–407.
39. Gould, S.M. Potential use of classroom response systems (CRS, Clickers) in foods, nutrition, and dietetics higher education. *J. Nutr. Educ. Behav.* **2016**, *48*, 669–674. [CrossRef]
40. Katz, L.; Hallam, M.C.; Duvall, M.M.; Polsky, Z. Considerations for using personal Wi-Fi enabled devices as "clickers" in a large university class. *Act. Learn. High. Educ.* **2017**, *18*, 25–35. [CrossRef]
41. Bryson, C. Engagement through partnership: Students as partners in learning and teaching in higher education. *Int. J. Acad. Dev.* **2016**, *21*, 84–86. [CrossRef]
42. Christie, M.; de Graaff, E. The philosophical and pedagogical underpinnings of active learning in engineering education. *Eur. J. Eng. Educ.* **2017**, *42*, 5–16. [CrossRef]
43. Chiu, P.H.P. A technology-enriched active learning space for a new gateway education programme in Hong Kong: A platform for nurturing student innovations. *J. Learn. Spaces* **2016**, *5*, 52–60.
44. Dreher, R.; Simpson, C.; Sørensen, O.J.; Turcan, R.V. (Eds.) When Students Take the Lead: Enhancing Quality and Relevance of Higher Education through Innovation in Student-Centred Problem-Based Active Learning. In Proceedings of the PBLMD International Conference, Chisinau, Moldova, 27–28 October 2016.
45. Englund, C.; Olofsson, A.D.; Price, L. Teaching with technology in higher education: Understanding conceptual change and development in practice. *High. Educ. Res. Dev.* **2017**, *36*, 73–87. [CrossRef]
46. Hassan, N.F.; Saifullizam, P. A survey of technology enabled active learning in teaching and learning practices to enhance the quality of engineering students. *Adv. Sci. Lett.* **2017**, *23*, 1104–1108. [CrossRef]
47. Jones, C.; Shao, B. *The Net Generation and Digital Natives, Implications for Higher Education*; A Literature Review; Higher Education Academy: Milton Keynes, UK, 2011; p. 56.
48. Hunsu, N.J.; Adesope, O.; Bayly, D.J. A meta-analysis of the effects of audience response systems (clicker-based technologies) on cognition and affect. *Comput. Educ.* **2016**, *94*, 102–119. [CrossRef]
49. Shapiro, A.M.; Sims-Knight, J.; O'Rielly, G.V.; Capaldo, P.; Pedlow, T.; Gordon, L.; Monteiro, K. Clickers can promote fact retention but impede conceptual understanding: The effect of the interaction between clicker use and pedagogy on learning. *Comput. Educ.* **2017**, *111*, 44–59. [CrossRef]
50. Hassanin, H.; Essa, K.; El-sayed, M.A.; Attallah, M.M. Enhancement of student learning and feedback of large group engineering lectures using audience response systems. *J. Mater. Educ.* **2016**, *38*, 175–190.
51. Selvi, R.T.; Chandramohan, G. Peer Assessment of Oral Presentation: An Investigative Study of Using Clickers in First-Year Civil Engineering Class of a Reputed Engineering Institution. In Proceedings of the 2016 IEEE Eighth International Conference on Technology for Education (T4E), Mumbai, India, 2–4 December 2016; pp. 132–135.
52. Khan, P.; O'Rourke, K. *Guide to Curriculum Design: Enquiry-Based Learning*; Higher Education Academy: Manchester, UK, 2004.
53. Kolb, D.A. *Experiential Learning: Experience as the Source of Learning and Development*; Pearson Education: Englewood Cliffs, NJ, USA, 2015.
54. Eyler, J. The power of experiential education. *Lib. Educ.* **2009**, *95*, 24–31.
55. Kaewunruen, S. *Final Report for Effective Academic Practice in Higher Education, PGCert in Academic Practice*; The University of Birmingham: Birmingham, UK, 2016; p. 24.
56. Conner, H.; Dench, S.; Bates, P. *An Assessment of Skill Needs in Engineering*; Department for Education and Employment Publications: Nottingham, UK, 2000.
57. Zaharim, A.; Md Yusoff, Y.; Omar, M.Z.; Mohamed, A.; Muhamad, N. Engineering Employability Skills Required by Employers in Asia. In Proceedings of the 6th WSEAS International Conference on Engineering Education, Rodos Island, Greece, 22–24 July 2009.

58. Hoekstra, A.; Mollborn, S. How clicker use facilitates existing pedagogical practices in higher education: Data from interdisciplinary research on student response systems. *Learn. Media Technol.* **2012**, *37*, 303–320. [CrossRef]

59. Camacho-Miñano, M.D.; del Campo, C. Useful interactive teaching tool for learning: Clickers in higher education. *Interact. Learn. Environ.* **2016**, *24*, 706–723. [CrossRef]

60. McLinden, M.; Hinton, D. *EBL for teachers of the visually impaired, Talking about Learning & Teaching Case Study 008*; University of Birmingham: Birmingham, UK, 2010.

61. Vuopala, E.; Hyvonen, P.; Jarvela, S. Interaction forms in successful collaborative learning in virtual learning environments. *Act. Learn. High. Educ.* **2016**, *17*, 25–38. [CrossRef]

62. Kaewunruen, S.; Tang, T. Idealisations of Dynamic Modelling for Railway Ballast in Flood Conditions. *Appl. Sci.* **2019**, *9*, 1785. [CrossRef]

63. Setsobhonkul, S.; Kaewunruen, S.; Sussman, J.M. Lifecycle Assessments of Railway Bridge Transitions Exposed to Extreme Climate Events. *Front. Built Environ.* **2017**, *3*, 35. [CrossRef]

64. Kaewunruen, S.; Sussman, J.M.; Matsumoto, A. Grand Challenges in Transportation and Transit Systems. *Front. Built Environ.* **2016**, *2*, 4. [CrossRef]

© 2019 by the author. Licensee MDPI, Basel, Switzerland. This article is an open access article distributed under the terms and conditions of the Creative Commons Attribution (CC BY) license (http://creativecommons.org/licenses/by/4.0/).

education sciences

MDPI

Article

Lessons Learned from the Development of Open Educational Resources at Post-Secondary Level in the Field of Environmental Modelling: An Exploratory Study

Quazi K. Hassan [1,*], Khan R. Rahaman [2], Kazi Z. Sumon [3] and Ashraf Dewan [4]

1 Department of Geomatics Engineering, Schulich School of Engineering, University of Calgary, 2500 University Dr. NW, Calgary, AB T2N 1N4, Canada
2 School of Urban and Regional Planning, University of Alberta, 116 St. and 85 Ave., Edmonton, AB T6G 2R3, Canada; krahaman@ualberta.ca
3 Department of Chemical and Petroleum Engineering, Schulich School of Engineering, University of Calgary, 2500 University Dr. NW, Calgary, AB T2N 1N4, Canada; kazi.sumon@ucalgary.ca
4 Spatial Sciences Discipline, School of Earth and Planetary Sciences, Curtin University, Kent St, Bentley, WA 6102, Australia; a.dewan@curtin.edu.au
* Correspondence: qhassan@ucalgary.ca; Tel.: +1-403-210-9494

Received: 29 March 2019; Accepted: 6 May 2019; Published: 13 May 2019

check for updates

Abstract: Open educational resources (OER) have become increasingly popular in recent times. Here, the aim was to synthesise the lessons learned through the development of OER materials for a university-level course called "environmental modelling". Consequently, the topics of discussion included: (i) how to choose an appropriate creative commons license; (ii) ways of incorporating materials from other sources, such as publicly available sources, other open access materials, and an author's own published materials if not published under a creative commons license; (iii) the impact of the developed OER in the field of environmental modelling; and (iv) the challenges in developing OER material. Upon developing the materials, we observed the following: (i) students enrolled in the course did not purchase textbooks; (ii) our OER materials ranked as one of the most accessed (i.e., number 7) materials according to the usage data that summed the number of file downloads and item views from PRISM (i.e., the hosting platform maintained by the University of Calgary); (iii) the students learned relatively better as per the data acquired by the University of Calgary's universal student ratings of instruction (USRI) instrument; and (iv) other universities expressed interest in adopting the materials.

Keywords: creative commons license; environmental educational materials; OER developmental challenges; OER adaptation; OER adaptation; OER creation

1. Introduction

Open education resources (OER) are materials that are openly available for reuse, modification, and sharing that originated in the later part of the twentieth century [1]. In 1994, Wayne Hodgins introduced the term "learning object" to describe any packaged digital resource useful for educational purposes [2,3]. Consequently, the concept of generative learning objects (GLOs) was developed to overcome restrictions that made it difficult to gain full benefits from the use of these generated resources [4]. The term "generative" is understood as a property of the learning content produced and handled either semi-automatically or automatically [5] with the support of some technology. As a result, GLOs have been considered as a leading technology of choice for eLearning support due to their potential generativity, adaptability, and scalability [6], as computer-based teaching components are considered an inevitable part of teaching and learning at post-secondary institutions. Many eLearning resources have been developed for specific environments, which restricts them from being in others. Due to this, in order to gain a positive return on investment, it is crucial that these eLearning resources are able to be repurposed, reused, and to be useable on a number of different platforms [7]. Usually a GLO has five major stages: (i) orientation stage; (ii) disclosure of ideas stage; (iii) challenges and restructuring stage; (iv) implementation stage; and (v) looking back stage [8]. In most cases, it is difficult to follow all the mentioned stages. Moreover, the idea has been criticized as it does not include options for the reuse and adaptation of developed materials in light of the pedagogical richness of the learning objects. Additionally, in order to overcome the limitations of the traditional ways of classroom teaching, computer games and interactive engagement through computer simulations have been gaining popularity [9,10]. However, the challenges remain in creating new materials that consider the licensing systems with appropriate permissions of materials used in these resources.

As a matter of fact, the term OER was first widely introduced at a United Nations Educational, Scientific, and Cultural Organization (UNESCO) conference in 2000, and it was promoted in the context of providing free access to educational resources on a global scale [11]. Consequently, the wave of OER started to spread when the Massachusetts Institute of Technology (MIT) OpenCourseWare [12] and the Organization for Economic Co-operation and Development (OECD) [13] took initiatives to develop OER in the early twenty-first century. One of the most commonly accredited definitions for the term OER then came along from the OECD report as "digested materials offered freely and openly for educators, students, and self-learners to use and reuse for teaching, learning, and research" [13]. It is worth noting that the resources in the mentioned definition were not limited to content, but comprised three areas [13]:

(i)   Learning content: Full courses, courseware, content modules, learning objects, collections, and journals;
(ii) Tools: Software to support the development, use, reuse, and delivery of learning content and materials;
(iii) Implementation resources: Intellectual property licenses to promote open publishing of materials, design principles of best practices, and localize content.

Therefore, OER mainly comprise teaching and learning, software-based tools and services, and licenses that allow open development and reuse of content, tools, and services [14,15].

On a global scale, there is a demand for high-quality OER materials that support the educational resources accessible to stakeholders related to teaching and learning at different geographical locations [16–19]. In this context, it would be worth mentioning some of the prominent OER developments for university-level courses: (i) MIT OpenCourseWare [12] and Rice University's Connexions project [20] in the United States; (ii) BCcampus OpenEd in Canada [21]; and (iii) the Japanese Open Course Ware Consortium [22] that provides access to about 400 courses from its 19 member universities. Moreover, the OpenCourseWare consortium (one of the global networks for open education) is collaborating with more than 200 higher education institutions and associate organizations from around the world to create a broad and deep body of open educational content using a shared model [23]. The mission there was to advance education and empower people worldwide through open courseware. Very recently in 2017, the global health charity, Bill and Melinda Gates Foundation, partnered with the American Association for the Advancement of Science (AAAS) in a yearlong agreement to "expand access to high-quality scientific publishing", which means that the foundation's funded research can be published on open access terms in *Science* and four other relevant journals [19].

Canada is unique in terms of education, as the federal government has no authority in the education sector, and education is exclusively a provincial responsibility [24]. However, there is a federal program to promote the growth of the open data movement through new open data licensing when the federal government funds research projects [25]. The governments of the three western Canadian provinces (i.e., Alberta, British Columbia, and Saskatchewan) have initiated the process of generating OER for post-secondary education in recent years [26]. In Alberta, the University of Calgary (UofC) spent approximately $13.63 million to acquire academic resources in the 2015–2016 session that included electronic journals, databases, and other scholarly materials, according to UofC's Libraries and Cultural Resources [27]. Additionally, there is another interesting fact—academics, including the UofC community, produce vast amounts of knowledge and synthesize this in the form of books, peer-reviewed journal articles, conference proceeding papers, reports, etc. However, the publishers sell these to the users (including the UofC community) again and again in the event that we opt to use them to disseminate knowledge from time to time. In the face of the above obstacles, the Canadian Tri-Council consisting of the following three granting agencies: (i) Natural Sciences and Engineering Research Council (NSERC); (ii) Social Sciences and Humanities Research Council (SSHRC); and (iii) Canadian Institutes of Health Research (CIHR), has adopted policies that their grantees should publish in open access platforms so that the generated knowledge is freely available to all the users across the world. Similar steps have also been adopted by the governmental funding agencies of USA, UK, Germany, and Australia [14,16,28]. It is worth mentioning that UofC has established an open access author's fund to help stakeholders in the event that they have exhausted all other funding sources [29].

As the concept and activities of openness are clearly evident in many Canadian universities and community colleges (i.e., post-secondary institutions), to introduce OER materials in order to broaden open access and build learning repositories, we have opted to create course materials in the Schulich School of Engineering at the University of Calgary. In doing so, we have generated a series of lecture notes under the umbrella of OER for an environmental modelling course (the course is taught at both graduate and undergraduate level) [30]. These lecture notes are freely accessible through the repository of the University of Calgary called PRISM [31]. In the scope of this paper, the aim was to synthesize the lessons learned from the development of the above-mentioned materials. Specifically, we focused on the following four issues: (i) adoption of an appropriate creative commons license; (ii) ways of incorporating materials from other sources such as publicly available sources, other open access materials, and an author's own published materials if not published under a creative commons license; (iii) the impact of the developed materials in the field of environmental modelling, both in-house and globally; and (iv) the challenges in developing OER material.

2. Methods

Our proposed methods consisted of three major components. Those included the following: (i) choosing the appropriate license for the OER materials; (ii) ways of incorporating materials from various resources; and (iii) analysing the impact of the developed OER. All these components are briefly described in the following subsections.

2.1. Choosing an Appropriate License

As we intended to develop OER materials, we opted to select a Creative Commons (CC) license that would allow people to copy, adapt, and distribute materials without requesting permission from the resource creator or paying license fees [32]. These licenses do not conflict with the copyright principles; they are a modification to "all rights reserved" copyrights [33]. In the scope of CC licensing, one of the most widely used approaches is the selection of CC BY, which allows anyone who complies with the standard CC terms to use the work for any purpose they wish (e.g., commercial use). In addition, there are several variations of CC BY licensing. For example, Katz [34] assembled the CC BY licensing attributions into three categories: (i) CC BY-No Derivatives (ND), which does not allow the creation of derivative works; (ii) CC BY-Share Alike (SA), which grants full rights to derivative works but requires them to be released under the same license as the original work (if derivative works are distributed, displayed, or performed); and (iii) CC BY-Non Commercial (NC), which grants full rights to derivative works but does not allow them to be used for any commercial purposes.

Therefore, as academic researchers affiliated with post-secondary institutions, we opted to use the CC BY-NC-SA licensing system so that the generated OER materials could be restricted to use for academic and/or other non-commercial purposes and shared under the original terms and conditions. Similar measures were recommended by MIT OpenCourseWare [12] and the McGill model [11] so that the created/adopted/adapted OER materials could be of use for academic purposes for continuous support towards teaching and learning worldwide.

2.2. Ways of Incorporating Materials from Various Sources

Apart from creating the author-owned OER materials, we opted to adopt and/or adapt contents (e.g., mainly figures) from other sources as these would need fewer resources (i.e., human and/or financial). Those included the following: (i) existing CC BY licensed materials; (ii) public domain information/databases; and (iii) authors' own works where the copyright had already been transferred to a publisher. All these cases are briefly discussed in the following subsections.

2.2.1. Adopting/Adapting from CC BY Licensed Materials

If any materials were published under one of the CC BY licenses other than CC BY-ND, then we adopted/adapted it with appropriate attributions. Figure 1 is an example of the adoption of a graphic available from Ahmed et al. [35] that was originally published under a CC BY 4.0 license inside a research article. It depicts the wildland–urban interface (WUI), as shown by a dotted yellow line, along with buffer zones at 10 m, 30 m, 50 m, 70 m, and 100 m. In fact, panel (a) in Figure 1 shows an example of areas where forest/plants were removed to protect nearby communities from wildfire. On the other hand, panel (b) in Figure 1 shows an example of a wildland fire-induced vulnerable area, where the existence of the vegetation (reddish color) would favor the forest fire propagation into the communities. Additionally, we adopted many graphics available from Wikipedia, where we used the attribution as: ["Title of the graphics" by Author name from Wikipedia is licensed under license type [e.g., CC BY 3.0]; website].

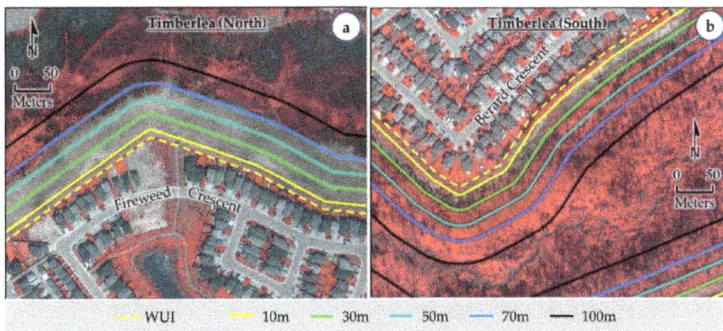

Figure 1. Example of the wildland–urban interface (WUI) and buffer zones around communities of interest that was "adopted from ©Ahmed, Rahaman, & Hassan, 2015 and licensed under CC BY 4.0". Panel (**a**) illustrates an area where authorities removed vegetation from the buffer zones during the fire-event to protect a nearby community; and Panel (**b**) shows an example of wildland fire-induced vulnerable area.

Figure 2 illustrates the concept of modifying a graphic that was originally published in Hazaymeh and Hassan [36] under a CC BY 4.0 license in a research article. It described the methods of generating synthetic land surface temperature (LST) at 30 m spatial resolution with 8 day intervals upon fusing 8 day composite of Moderate Resolution Imaging Spectroradiometer (MODIS)-derived LST at 1 km and Landsat-derived LST at 30 m acquired every 16 days.

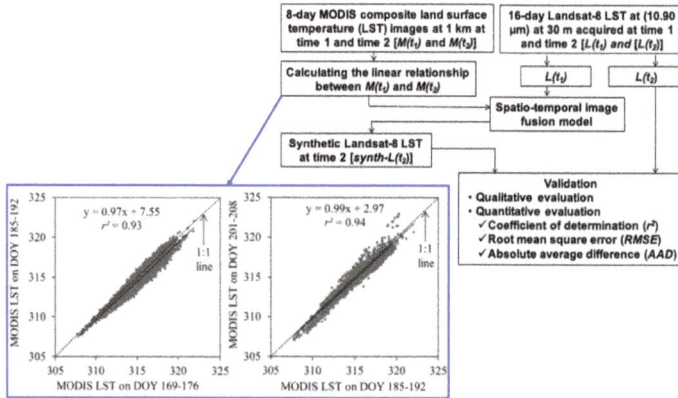

Figure 2. Example of down-scaling both of the spatial and temporal resolution of the satellite data that was "adapted from ©Hazaymeh, & Hassan, 2015 licensed under CC BY 4.0".

2.2.2. Adopting/Adapting Materials from the Public Domain

In case of developing OER materials, we often incorporated content available in the public domain, including US and Canadian government works. Note that most of the US government creative works including writing and images are not copyrighted [37]. However, such adoption/adaptation requires proper acknowledgement to the agency that has produced the work of interest [38]. In addition, it is clear that the Government of Canada permits the use of its work for non-commercial reproduction purposes [39], and such an illustration is provided in Figure 3. Note that forest occupies about 347 million hectares of Canada's landmass, which is about 9% of the global forest [40].

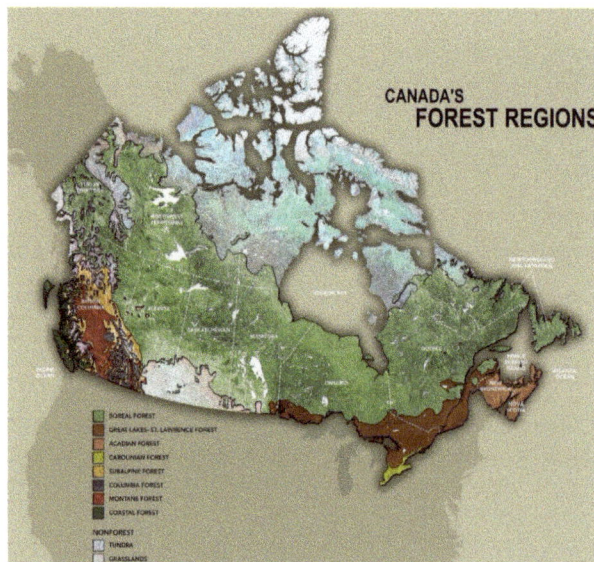

Figure 3. Canada's forest regions. Adopted from Natural Resources Canada that allows non-commercial reproduction. Such reproduction is a copy of an official work that is published by the Government of Canada, and the reproduction has not been produced in affiliation with, or with the endorsement of the Government of Canada. Available online: http://www.nrcan.gc.ca/forests/measuring-reporting/classification/13179 (accessed on 21 February 2019) [41].

2.2.3. Incorporating Authors' Own Works Published under Non-CC BY Licenses

In general, we, as authors, transfer the copyright of an article of interest to the publisher upon its acceptance (e.g., [42]). Thus, we must seek permission from the respective publisher prior to incorporating these articles into new/adapted work. To the best of our knowledge and experience, we find that most publishers are willing to allow authors to incorporate their own transferred copyright protected materials into new ones for non-commercial purposes. Such an example is shown in Figure 4, where the first author of this manuscript was the lead author of a manuscript published by the Canadian Aeronautics and Space Institute (CASI).

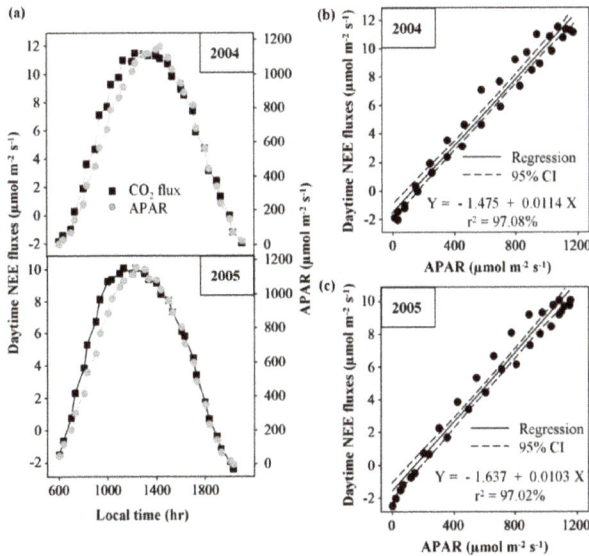

Figure 4. Relationship between absorbed photosynthetically active radiation (APAR) and daytime net ecosystem exchanges (NEE) over a balsam fir forest in New Brunswick, Canada. Panel (**a**) shows averaged May–September period daytime NEE as a function of local time for 2004 and 2005. Panel (**b**,**c**) illustrates linear fits applied to plots of daytime NEE as a function of APAR. Note that this graphic is "adopted from Hassan et al. [43] ©2006 CASI, where the publisher allows Hassan as an author to reproduce for non-commercial purposes".

2.3. Analysing the Impact of the Developed OER

We analysed the impact of the developed OER materials in four ways: (i) the potential cost savings for the students, (ii) usage statistics of the OER in the online repository, (iii) the impact on the students' learning using the University of Calgary's universal student ratings of instruction (USRI) instrument [44], and (iv) the expression of interest in adopting the materials by other universities. These are briefly described in Section 3—"Results and Discussion".

3. Results and Discussion

3.1. Brief Description of the Developed OER

We developed OER materials for the course called ENGO 583/ENEN 635 Environmental Modelling. It is a cross-listed course between geomatics engineering at a fourth year level (i.e., ENGO 583) and environmental engineering at a graduate level (ENEN 635). Figure 5 shows the screenshot of the home page of the developed OER materials available in the PRISM repository.

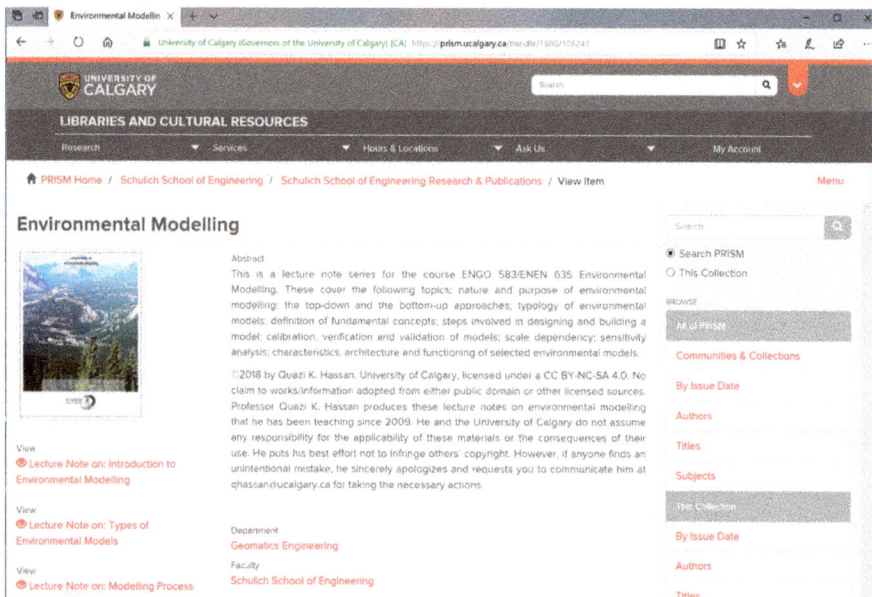

Figure 5. Home page of the developed OER (doi: http://dx.doi.org/10.11575/PRISM/5245).

The developed materials consisted of 20 Microsoft PowerPoint-based pdf files that are individually downloadable from PRISM. These materials were clustered into two broader themes or sections—"fundamentals of environmental modelling", and "modelling of environmental issues". The section "fundamentals of environmental modelling" consisted of seven lecture notes that discussed the following rudimentary concepts: (i) introduction to environmental modelling, (ii) types of environmental modelling, (iii) modelling processes, (iv) model sensitivity analysis, (v) model calibration and uncertainty analysis, (vi) model validation, and (vii) scaling issues in environmental modelling. The other section, "modelling of environmental issues", comprised 13 lecture notes that illustrated the mechanism of developing various models to discuss a set of environmental issues. These included: (i) solar radiation/energy modelling, (ii) evapotranspiration modelling, (iii) modelling the effect of greenhouse gases on temperature, (iv) wind energy modelling, (v) modelling potential species distribution, (vi) bioenergy modelling, (vii) forecasting forest fire danger conditions, (viii) modelling local warming trends, (ix) forecasting river water flow, (x) modelling vegetation phenological stages, (xi) forecasting rice yield, (xii) surface water quality modelling, and (xiii) mediated modelling.

3.2. Potential Cost Savings for the Students

We made our OER materials available during the winter 2018 semester, spanning between the months of January and April, and they were considered as the textbook for the students. To our knowledge, none of the students enrolled on the course bought any textbooks in addition to acquiring the generated OER materials. Consequently, we considered that every student saved at least $200 by avoiding the purchase of textbooks. The assumption of such cost lay in the fact that a student would be required to buy at least two textbooks to cover the material delivered in the scope of the course. It is worth noting that similar findings of cost savings by students or those not keen to purchase textbooks were reported in the literature for OER materials that are available on the internet (e.g., [45,46]).

3.3. Usage Statistics as Observed through the Online Repository

We acquired some quantitative data (all time usage and views/downloads in the past six months) available from the PRISM website that showed the usage and interest of the material worldwide. Since the launch of the material (i.e., January 2018), a total of 5605 downloads and 2353 views were recorded as of 22 April 2019, as shown in Figure 6. The top five countries either by downloads or by views in the past six months are also given in Figure 6 to demonstrate the education outreach of this particular OER material. In addition, these materials were found to be one of the most popular items (i.e., ranked number 7) after summing the number of file downloads and item views since the launch, as calculated on 22 April 2019 according to PRISM repository data. Thus, we assumed that the present courseware has created opportunities for potential cost savings for students and would continue to help in the learning of the general audience in this area worldwide.

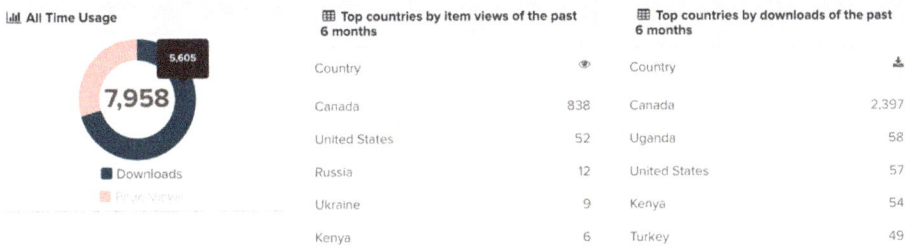

📊 All Time Usage		⊞ Top countries by item views of the past 6 months		⊞ Top countries by downloads of the past 6 months	
		Country	👁	Country	📥
	5,605	Canada	838	Canada	2,397
7,958		United States	52	Uganda	58
		Russia	12	United States	57
■ Downloads		Ukraine	9	Kenya	54
■ Page Views		Kenya	6	Turkey	49

Figure 6. Global usage and interest in the developed OER, as acquired on 22 April 2019 from PRISM.

In addition, we attempted to quantify the usage of the developed materials by analyzing the students enrolled in the course ENGO 583/ENEN 635. In fact, we used these materials in winter 2018 and 2019, when there were 42 enrolled students. If we assume that each student accessed these materials twice through both downloading and viewing, then the total count would be 1680 (= 42 (students) * 20 (number of lecture notes) * 2 (number of times accessed)). Thus, the remaining count (i.e., greater than 6000, as depicted in Figure 6) of the accessed materials might have been due to other learners across Canada and elsewhere in the world. Consequently, we could consider that the developed OER materials were able to generate considerable interest worldwide.

3.4. Impact of the OER on the Students' Learning

In order to assess the impact of the developed OER on the students' learning, we employed the universal student ratings of instruction (USRI) instrument, which has been in operation at the University of Calgary for at least two decades. Note that the USRI consists of 12 measures [47]; however, we analysed three of the measures, i.e., (i) content well organized; (ii) I [student] learned a lot in this course; and (iii) support materials helpful) that were related to the students' learning over the years 2014–2018. During this period, for the first three years, the course used traditional materials (which included materials extracted from non-OER sources), while we used the developed OER materials in 2018. Figure 7 shows the impact of the developed OER in comparison to the traditional materials on student learning. It revealed that the OER materials enhanced the students' learning ability in all three measures.

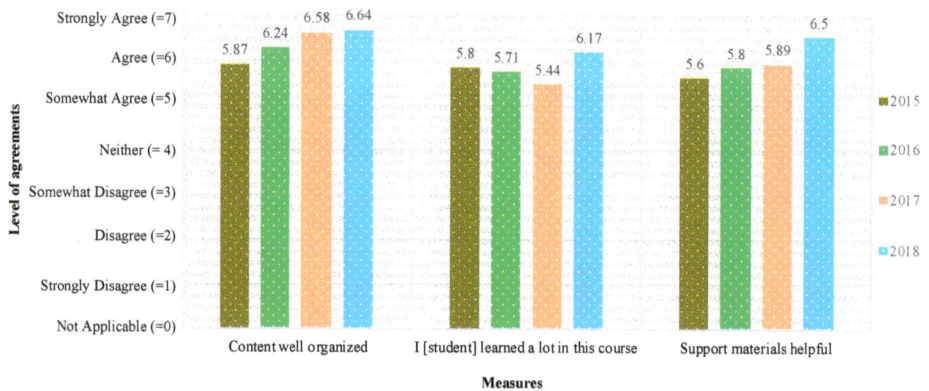

Figure 7. Impact of the developed OER on the students' learning in comparison to traditional materials analysed using some of the relevant measures acquired through the universal student ratings of instruction (USRI) instrument.

3.5. Expression of Interest in Adopting the Materials from Other Universities

Since the launch of these specific OER materials, we received several emails that expressed interest in adopting/adapting the generated materials, and these are summarized in Table 1. In addition, the second author of this article had already adapted some of the developed OER materials into an undergraduate course called "natural hazard modelling" at the University of Alberta. He would also like to do the same for another undergraduate course (i.e., remote sensing) in fall 2019 at the same institute.

Table 1. Higher academic institutes that showed interest in adopting/adapting our OER materials.

Ref.	Name of the Higher Educational Institute	Comments
[48]	Department of Earth and Environmental Sciences, Yarmouk University, Jordan.	Interested in adopting as a graduate course.
[49]	Department of Civil Engineering, COMSATS University-Abbottabad Campus, Pakistan	Indicated the usefulness and relevance of the OER in two undergraduate courses, i.e., "Geo-informatics" and "Disaster Management".
[50]	Department of Computer Science and Engineering, Khulna University of Engineering and Technology, Bangladesh.	Keen to adapt for teaching computing algorithms for environments.
[51]	BMS School of Architecture, Bangalore, India.	Interested in adapting for courses related to landscape and environment.
[52]	Department of Surveying and GeoInformatics, Akanu Ibiam Federal Polytechnic, Nigeria.	Sought permission to use for two graduate courses, i.e., "Environmental Applications of Remote Sensing" and "Land Use Planning".

3.6. Challenges of Developing OER

Post-secondary institutions around the world have been using the internet and other relevant technologies to develop and distribute teaching and learning materials for decades. OER have gained critical attention from stakeholders in teaching and learning communities for their potential and promise to obviate demographic, economic, and geographic boundaries of scholarships and education to promote life-long and personalized learning. The rapid growth of OER materials certainly provides new opportunities for teaching and learning; however, the challenges remain evident in several areas, including but not limited to the following:

- Scholars or creators of OER materials sometimes lack knowledge in assessing the proper compatibilities of licensing and attribution while adopting, adapting, or creating OER materials without any restrictions to be utilized immediately after hosting in repositories. We have identified this as a major challenge in generating OER materials because the created documents are already distributed in diverse geographical locations;
- Often, a lack of excitement and motivation from fellow colleagues and peers do not encourage scholars to create new knowledge in OER materials, although some adoption and adaptation may be evident. Additionally, immediate rewards are not discernible if a scholar develops OER materials to support his/her courses at post-secondary institutions;
- While generating our OER materials for hosting in the PRISM repository at the UofC, we identified that knowledge of OER licensing systems and compatibility checks is lacking on campus, and we recognized the lack of financial resources that support the adaptation (in particular, graphics) and/or generation of new materials.
- We came across an interesting challenge regarding the familiarity of OER materials among the student groups within the institution where we generated the materials. As a result, we assume, if the students are not well aware of the quality of OER materials, the generated content may not bring immediate impact to the learning communities. Note that an OER advocacy group was visible on campus to encourage adopting OER materials in the learning environment at the UofC, which may not be the case at other post-secondary institutions;
- Researchers have argued that the sustainability [53] and quality [28] of OER materials remain an important issue. However, after generating the OER materials, we received appreciation (see Section 3.5—"Expression of interest in adopting the materials from other universities" for details), and institutions from diverse developing countries are adopting these materials to teach in both graduate and undergraduate courses. In this circumstance, we assume that the additional challenge of creating OER is to consider the appropriateness of the material for the learner communities and if the material is sustainable and brings positive impacts to scholarly communities.

4. Concluding Remarks

In the scope of this paper, we summarized the lessons learned through the development of OER course materials for the field of environmental modelling. The major obstacle we encountered was the existence of limited OER that we could either adopt or adapt. On the other hand, we found an enormous number of figures available in the public domain, from places such as the following: (i) US Governmental organizations/departments, e.g., NASA, USGS, EPA, etc.; (ii) the Government of Canada; (iii) Wikipedia; (iv) Wikimedia; and (v) Flicker. In addition, we found that it would be possible to incorporate an authors' own work published under a non-CC BY license only if it was reproduced for non-commercial purposes. In this case, we suggest that a researcher consult the publisher prior to implementing this particular idea. Moreover, we evaluated the impact of the developed OER in four ways: (i) the students enrolled in the course called ENGO 583/ENEN 635 in winter 2018 did not purchase textbooks, thus saved the associated cost; (ii) the materials were either viewed or downloaded across the world, and our materials were seen to have a rank of number 7 in terms of the sum of file downloads and item views according to PRISM as of 22 April 2019; (iii) the students learned better according to the data acquired by the University of Calgary's USRI instrument, and (iv) at least five faculty members from various countries sought permission to adopt the materials for their relevant graduate/undergraduate courses. Despite this, we identified a set of challenges, such as the following: (i) scholars/creators' lack of knowledge about the licensing issues, (ii) lack of excitement and motivation from both the academics and students, and (iii) resources, particularly financial ones, are quite limited in the case of generating new OER. Finally, we believe that the lessons we learned through this exercise could encourage other academics across the world to generate new OER for facilitating the improvement of the teaching and learning environment.

Author Contributions: Conceptualization, Q.K.H., K.R.R.; writing—original draft preparation, Q.K.H., K.R.R., K.Z.S., and A.D.; writing—review and editing, Q.K.H., K.R.R., K.Z.S., and A.D.; supervision, Q.K.H.

Acknowledgments: The authors would like to thank the Office of Institutional Analysis at the University of Calgary for acquiring and providing the USRI data.

Conflicts of Interest: The authors declare no conflict of interest.

References

1. Hocevar, C.; Lange, J. *Challenges and Opportunities: Open Educational Resources (OERs) at McGill University*; Student Society of McGill University: Montreal, QC, Canada, 2017; 73p.
2. Mossley, D.D. *Open Educational Resources and Open Education*; The Higher Education Academy: Heslington, UK, 2013; Available online: https://ils.unc.edu/courses/2014_fall/inls690_109/Readings/Mossely2013-OERandOpenEducation.pdf (accessed on 28 March 2019).
3. Koroivulaono, T. Open educational resources: A regional university's journey. *Int. J. Educ. Technol. High. Educ.* **2015**, *11*, 1–17. [CrossRef]
4. Oldfield, J.D. An implementation of the generative learning object model in accounting. In Proceedings of the 2008 Australasian Society for Computers in Learning in Tertiary Education, Melbourne, Australia, 30 November–3 Deccember 2008; ASCILITE: Melbourne, Australia; pp. 687–695.
5. Burbaite, R.; Bespalova, K.; Damaševičus, R.; Štuikys, V. Context-Aware Generative Learning Objects for Teaching Computer Science. *Int. J. Eng. Educ.* **2014**, *30*, 929–936.
6. Štuikys, V.; Damaševičus, R. Towards knowledge-based generative learning objects. *Inf. Technol. Control.* **2007**, *36*, 202–208.
7. Boyle, T. Design principles for authoring dynamic, reusable learning objects. *Australas. J. Educ. Technol.* **2016**, *19*, 46–58. [CrossRef]
8. Ariani, Y. Developing college students' soft skills through generative learning model. *Adv. Soc. Sci. Educ. Humanit. Res.* **2017**, *118*, 830–834.
9. Oyesiku, D.; Adewumi, A.; Misra, S.; Ahuja, R.; Damaševičus, R.; Maskeliunas, R. An Educational Math Game for High School Students in Sub-Saharan Africa. *Commun. Comput. Inf. Sci.* **2018**, *942*, 228–238.
10. Raziunaite, P.; Miliunaite, A.; Maskeliunas, R.; Damasevicius, R.; Sidekerskiene, T.; Narkeviciene, B. Designing an educational music game for digital game based learning: A Lithuanian case study. In Proceedings of the 2018 41st International Convention on Information and Communication Technology, Electronics and Microelectronics, Opatija, Croatia, 21–25 May 2018; pp. 800–805.
11. Glennie, J.; Harley, K.; Butcher, N.; van Wyk, T. (Eds.) *Open Educational Resources and Change in Higher Education: Reflections from Practice*; Commonwealth of Learning: Vancouver, BC, Canada, 2012; 291p.
12. 2005 Program Evaluation Findings Report: MIT OpenCourseWare. Available online: https://ocw.mit.edu/ans7870/global/05_Prog_Eval_Report_Final.pdf (accessed on 28 March 2019).
13. Giving Knowledge for Free: The Emergence of Open Educational Resources. Available online: http://www.oecd.org/education/ceri/38654317.pdf (accessed on 28 March 2019).
14. Santos, A.; McAndrew, P.; Godwin, S. Open educational resources: New directions for technology-enhanced distance learning in the third millenium. *Formamente* **2008**, *1–2*, 111–126.
15. Picasso, V.; Phelan, L. The evolution of open access to research and data in Australian higher education. *Int. J. Educ. Technol. High. Educ.* **2015**, *11*, 122–133. [CrossRef]
16. Richter, T.; McPherson, M. Open educational resources: Education for the world? *Distance Educ.* **2012**, *33*, 201–219. [CrossRef]
17. Stacey, P. Government support for open educational resources: Policy, funding, and strategies. *Int. Rev. Res. Open Distance Learn.* **2013**, *14*, 67–80. [CrossRef]
18. Garza, L.; Sancho-Vinuesa, T.; Zermeno, M. Indicators of pedagogical quality for the design of a massive open online course for teacher learning. *Int. J. Educ. Technol. High. Educ.* **2015**, *12*, 104–118.

19. Van Noorden, R. Science journals end open-access trial with Gates Foundation. *Nature* **2018**, *559*, 311–312. [CrossRef] [PubMed]

20. Rice University's Connexions Project Pioneers Open-Source Academic Publishing | EurekaAlert! Science News. Available online: https://www.eurekalert.org/pub_releases/2004-02/ru-wsl021904.php (accessed on 28 March 2019).

21. BCcampus OpenEd Resources—Learning about, and Experiencing, Open Educational Practices. Available online: https://open.bccampus.ca (accessed on 24 April 2019).

22. Japan Open Course Ware Consortium: A Case Study in Open Educational Resources Production and Use in Higher Education. Available online: http://www.oecd.org/education/ceri/37647892.pdf (accessed on 28 March 2019).

23. Butcher, N. Open Educational Resources and Higher Education. Available online: http://oerworkshop.weebly.com/uploads/4/1/3/4/4134458/03_open_educational_resources_and_higher_education.pdf (accessed on 28 March 2019).

24. Council of Ministers of Education, Canada. Available online: https://www.cmec.ca/299/education-in-canada-an-overview/index.html (accessed on 28 March 2019).

25. Scassa, T. Canada's New Draft Open Government Licence. Available online: http://www.teresascassa.ca/index.php?option=com_k2&view=item&id=113:canadas-new-draft-open-government-licence&Itemid=83 (accessed on 28 March 2019).

26. McGreal, R.; Anderson, T.; Conrad, D. Open educational resources in Canada 2015. *Int. Rev. Res. Open Distance Learn.* **2015**, *16*, 161–175. [CrossRef]

27. Budget—Content Development—Library at University of Calgary. Available online: https://library.ucalgary.ca/c.php?g=255277&p=3804033 (accessed on 28 March 2019).

28. Yuan, L.; Macneill, S.; Kraan, W. Open Educational Resources-Opportunities and Challenges for Higher Education. Available online: http://publications.cetis.org.uk/wp-content/uploads/2012/01/OER_Briefing_Paper_CETIS.pdf (accessed on 28 March 2019).

29. Open Access Authors Fund—Scholarly Communication—Library at University of Calgary. Available online: https://library.ucalgary.ca/guides/scholarlycommunication/oafund (accessed on 28 March 2019).

30. Hassan, Q.K. *Environmental Modelling*; University of Calgary: Calgary, AB, Canada, 2018; Available online: http://dx.doi.org/10.11575/PRISM/5245 (accessed on 28 March 2019).

31. PRISM Home. Available online: https://prism.ucalgary.ca/ (accessed on 28 March 2019).

32. Levey, L. Finding relevant OER in higher education: A personal account. In *Open Educational Resources and Change in Higher Education: Reflections from Practice*; Glennie, J., Harley, K., Butcher, N., van Wyk, T., Eds.; Commonwealth of Learning: Vancouver, BC, Canada, 2012; pp. 125–140.

33. De los Arcos, B.; Farrow, R.; Perryman, L.-A.; Pitt, R.; Weller, M. *OER Evidence Report 2013–2014.* Open Research Hub. 2014. Available online: http://oro.open.ac.uk/41866/1/oerrh-evidence-report-2014.pdf (accessed on 28 March 2019).

34. Katz, Z. Pitfalls of open licensing: An analysis of creative commons licensing. *IDEA* **2006**, *46*, 391–413.

35. Ahmed, M.R.; Rahaman, K.R.; Hassan, Q.K. Remote sensing of wildland fire-induced risk assessment at the community level. *Sensors* **2018**, *18*, 1570. [CrossRef] [PubMed]

36. Hazaymeh, K.; Hassan, Q.K. Fusion of MODIS and Landsat-8 surface temperature images: A new approach. *PLoS ONE* **2015**, *10*, e0117755. [CrossRef] [PubMed]

37. U.S. Government Works, USAGov. Available online: https://www.usa.gov/government-works (accessed on 28 March 2019).

38. NASA Image Use Policy, Precipitation Measurement Missions. Available online: https://pmm.nasa.gov/image-use-policy (accessed on 28 March 2019).

39. Terms and Conditions, Natural Resources Canada. Available online: https://www.nrcan.gc.ca/terms-conditions/10847 (accessed on 28 March 2019).

40. How Much Forest Does Canada Have? | Natural Resources Canada. Available online: https://www.nrcan.gc.ca/forests/report/area/17601 (accessed on 28 March 2019).

41. Forest Classification, Natural Resources Canada. Available online: https://www.nrcan.gc.ca/forests/measuring-reporting/classification/13179 (accessed on 28 March 2019).

42. American Scientific Publishers: Journal of Biobased Materials and Bioenergy—Copyright Transfer Agreement. Available online: http://www.aspbs.com/jbmbe/JBMBE%20Copyright%20Transfer%20Form.pdf (accessed on 28 March 2019).

43. Hassan, Q.K.; Bourque, C.P.-A.; Meng, F.-R. Estimation of daytime net ecosystem CO2 exchange over balsam fir forests in eastern Canada: Combining averaged tower-based flux measurements with remotely sensed MODIS data. *Can. J. Remote Sens.* **2006**, *32*, 405–416. [CrossRef]

44. Universal Student Ratings of Instruction (USRI—Course Evaluation)—Office of Institutional Analysis at University of Calgary. Available online: https://www.ucalgary.ca/usri/ (accessed on 23 April 2019).

45. Bliss, T.J.; Hilton, J., III; Wiley, D.; Thanos, K. The cost and quality of online open textbooks: Perceptions of community college faculty and students. *First Monday* **2013**, *18*, 1. [CrossRef]

46. Weller, M.; de los Arcos, B.; Farrow, R.; Pitt, B.; McAndrew, P. The impact of OER on teaching and learning practice. *Open Praxis* **2015**, *7*, 351–361. [CrossRef]

47. Universal Student Ratings of Instruction (USRI—Course Evaluation)—Office of Institutional Analysis at University of Calgary. Available online: https://www.ucalgary.ca/usri/node/161 (accessed on 23 April 2019).

48. Al-Rawabdeh, A.; (Yarmouk University, Irbid, Jordan). Personal communication, 2018.

49. Akbar, T.A.; (COMSATS University- Abbottabad Campus, Abbottabad, Pakistan). Personal communication, 2018.

50. Rabbi, J.; (Khulna University of Engineering and Technology, Khulna, Bangladesh). Personal communication, 2019.

51. Cherian, R.; (BMS School of Architecture, Bengaluru, India). Personal communication, 2019.

52. Ejiaha, I.R.; (Akanu Ibiam Federal Polytechnic, Amoncha, Nigeria). Personal communication, 2019.

53. Atkins, D.E.; Brown, J.S.; Hammond, A.L. *A Review of the Open Educational Resources (OER) Movement: Achievements, Challenges, and New Opportunities*; A Report to The William and Flora Hewlett Foundation; 2007; 80p, Available online: https://hewlett.org/wp-content/uploads/2016/08/ReviewoftheOERMovement.pdf (accessed on 28 March 2019).

© 2019 by the authors. Licensee MDPI, Basel, Switzerland. This article is an open access article distributed under the terms and conditions of the Creative Commons Attribution (CC BY) license (http://creativecommons.org/licenses/by/4.0/).

education
sciences

MDPI

Article

Tutorials for Integrating CAD/CAM in Engineering Curricula

AMM Sharif Ullah [1],* and **Khalifa H. Harib [2],***

1   Faculty of Engineering, Kitami Institute of Technology, Kitami 090-8507, Japan
2   College of Engineering, United Arab Emirates University, Al Ain 15551, UAE
*   Correspondence: ullah@mail.kitami-it.ac.jp (A.S.U.); k.harib@uaeu.ac.ae (K.H.H.);
      Tel.: +81-157-26-9207 (A.S.U.)

Received: 23 August 2018; Accepted: 14 September 2018; Published: 19 September 2018

check for
updates

Abstract: This article addresses the issue of educating engineering students with the knowledge and skills of Computer-Aided Design and Manufacturing (CAD/CAM). In particular, three carefully designed tutorials—cutting tool offsetting, tool-path generation for freeform surfaces, and the integration of advanced machine tools (e.g., hexapod-based machine tools) with solid modeling—are described. The tutorials help students gain an in-depth understanding of how the CAD/CAM-relevant hardware devices and software packages work in real-life settings. At the same time, the tutorials help students achieve the following educational outcomes: (1) an ability to apply the knowledge of mathematics, science, and engineering; (2) an ability to design a system, component, or process to meet the desired needs, (3) an ability to identify, formulate, and solve engineering problems; and (4) an ability to use the techniques, skills, and modern engineering tools that are necessary for engineering practice. The tutorials can be modified for incorporating other contemporary issues (e.g., additive manufacturing, reverse engineering, and sustainable manufacturing), which can be delved into as a natural extension of this study.

Keywords: engineering education; CAD/CAM; digital engineering; accreditation

---

1. Introduction

Manufacturing is a wealth-generating activity, and, thereby helps ensure the economic well-being of a country. Nowadays, computer-aided hardware devices (e.g., Computer Numerically Controlled (CNC) machine tools, three-dimensional (3D) printers, robots, programmable logic controller, 3D scanners, Coordinate Measuring Machines, computerized assembly lines, and alike) and software packages for solid modeling (e.g., 3D Computer-Aided Design (CAD) and Computer-Aided Manufacturing (CAM) packages) are extensively used to perform the activities related to product life cycle (conceptualize, design, manufacture, assemble, distribute, use, maintain, and recycle). Therefore, the individuals who can design, analyze, use, and maintain the hardware devices and software packages mentioned above are essential for the economic growth of a country (in this article, the above-mentioned devices and software packages are collectively referred to as CAD/CAM). As such, higher education institutes all over the world need to educate students with the knowledge and skills of CAD/CAM and other relevant issues for meeting industry needs [1–3]. Thus, engineering degree programs all over the world offer courses related to CAD/CAM [4]. Similar to the courses where the students learn engineering science (e.g., [5]), a technology-oriented course also needs customized methods and tools for the educational effectiveness and outcomes assessment (e.g., see the tools developed for the courses related to CAD [6] and courses related to the Internet of Things (IoT) [7]). This is true for the education of CAD/CAM in the academia. Some of the selected articles that deal with the educational aspects of CAD/CAM in academia are described below. Araki [8] and Zhao et al. [9]

described the contents to educate learners to use the CAD/CAM packages, which are made available through the multimedia network. deWeck et al. [10] described the students' activities and their performances while using commercially available CAD/CAM systems and other relevant software packages. The purpose is to provide students with the opportunity to materialize a conceptual design. Ishitsuka et al. [11] described the methodology to teach CAD/CAM using an exercise for creating the CNC program from a given solid model prepared by using the CAD system. Lin et al. [12] described the teaching methodology to integrate CAD with CAM for undergraduate students. Alemzadeh [13] described how to introduce CAD/CAM from the perspective of concurrent design for senior-level engineering students. Kaneko and Kashimoto [14] developed a customized computer-aided tool for educating students with the knowledge and skills of CAD/CAM. Abouelala et al. [15] described a methodology for selecting a CAM package for educating students with the knowledge and skills of CAD/CAM, considering the opinions of the stakeholders. Moschonissiotis and Vosniakos [16] developed a methodology to train individuals with the knowledge and skills of CAD/CAM where a set of structured rules and utilization of these rules in process planning are emphasized for achieving tangible results.

However, consider the issue of the assessment of students' performance. In this regard, Main and Staines [17] reported that the practical examinations are a more informative and efficacious method for assessing the students' performance in CAD/CAM. Hamade et al. [18] reported how to evaluate the skills of the senior-level students while building solid models using commercially available CAD packages. In both cases, the learners are the mere users of the available technology. Whether or not the students are aware of how the systems work and why the systems can produce the desired shape or product has not been the issue of the assessments mentioned above. This is perhaps not the goal of the outcomes-oriented education programs recommended by the accreditation bodies (e.g., ABET, http://www.abet.org). The reason is that the educational activities underlie three aspects: namely, idealism, realism, and pragmatism. The idealism and realism prepare an individual to be a "knower", whereas pragmatism prepares an individual to be "doer" only. Therefore, several outcomes are set so that the student becomes a "knower" and "doer" simultaneously. As far as the ABET-recommended outcomes are concerned, the content of a course (or a series of courses) regarding CAD/CAM should help students attain at least the following four outcomes: (1) an ability to apply the knowledge of mathematics, science, and engineering; (2) an ability to design a system, component, or process to meet the desired needs; (3) an ability to identify, formulate, and solve engineering problems; and (4) an ability to use the techniques, skills, and modern engineering tools necessary for engineering practice. At the same time, the contents must be derived from the fundamental aspects of CAD/CAM, as schematically illustrated in Figure 1. In particular, the region defined by the dotted box in Figure 1 illustrates the central issues underlying CAM. As shown in Figure 1, in CAD/CAM, the CAD data of the object to be manufactured becomes the main input. Based on this input, the user sets the blank and the cutting tool to be used. The user then selects the appropriate tool-paths (morphing, zigzag, and/or contour parallel), the cutting tool parameters (number of cutting edges, tool nose radius, tool length, and/or tool diameter), and cutting operation (rough-cut and or finish cut). Based on the selections that the user made, the CAD/CAM package simulates the tool-paths, showing their effects on the blank. When the simulation result satisfies the user (i.e., a simulation result produces the desired shape), the CAD/CAM package creates the CNC program, i.e., it formats (i.e., post-processes) the cutter locations given by the tool-paths and the cutting conditions (feed rate, cutting velocity, depth of cut, and alike) in a way so that the controller of the CNC machine tool can understand and execute the program. Once the user is satisfied with the program, it can be executed to produce the desired shape or product, as schematically illustrated in Figure 1. The digital measurement technology can be used to report the dimensional and form accuracies, as well as the surface roughness for the sake of quality control.

Based on the above-mentioned description of the educational aspects of CAD/CAM, this article presents three tutorials. Thus, the remainder of this article is organized as follows: Section 2 describes the first tutorial for calculating the tool-path for machining a 2D shape given by a parametric curve. Section 3 describes the second tutorial for determining 3D tool-path for machining a freeform surface, which is given by a parametric surface. Section 4 describes the last tutorial, where one integrates the advanced machine tools (e.g., hexapod-based machine tools) and geometric modeling of freeform surfaces. Section 5 provides the concluding remarks of this study.

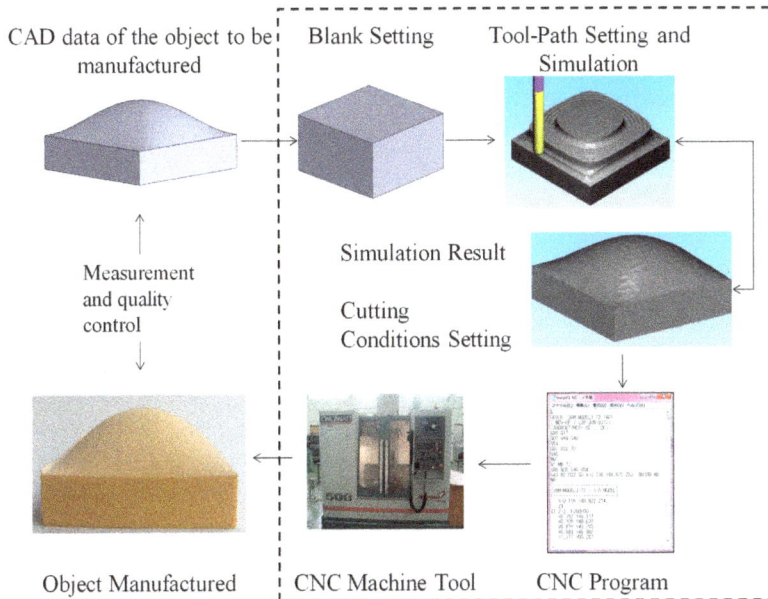

Figure 1. The flow diagram of Computer-Aided Design and Manufacturing (CAD/CAM).

2. Tutorial for 2D Tool-Path Generation

As shown in Figure 1, one of the major tasks of CAD/CAM is to create the tool-paths. How do the CAD/CAM packages generate the tool-paths? The tutorial that helps a student understand the process of creating the tool-paths can also help achieve the outcomes (1)–(4) as mentioned above. According, this section describes a tutorial for generating a two-dimensional (2D) tool-path. For the sake of better understanding, the parametric curve-based tool offsetting method is used in the tutorial. The description is as follows:

Parametric curves are extensively used to represent a large variety of engineering shapes and parts [19,20]. The following equation defines a parametric curve denoted as $C(t)$.

$$C(t) = \sum_{i=1}^{n} P_i \times B_i(t) = [t][M][G] \tag{1}$$

In Equation (1), P_i is the i-th control point, $B_i(t)$ is the blending function associated with P_i, and t is the parameter so that $\forall\, t \in [0,1]$. As such, the matrix that is denoted as $[t]$, which is a row matrix, is called the parameter matrix. The matrix that is denoted as $[M]$, which is an $n \times n$ matrix, is called the shape matrix. The matrix that is denoted as $[G]$, which is a column matrix, is called the control point, vertex, or geometric matrix.

When one models a shape using a parametric curve as defined in Equation (1), the user must decide the number of control points (n), the set of control points $\{P_i \mid i = 1,...,n\}$, and the blending functions $\{B_i \mid i = 1,...,n\}$. The blending functions depend on both the number of control points and the type of parametric curve (e.g., Bezier curve, B-spline, and alike). The number of control points and the type of the parametric curve decide [M] and [t]. For example, consider the case of a parametric curve called the quadratic B-spline ($C_{QB}(t)$). The coordinates of the three arbitrary control points (P_{p1}, P_{p2}, and P_{p3}) are needed to construct the [G] for $C_{QB}(t)$. Here, the subscript "p" denotes one of the coordinates out of x, y, and z, i.e., $p \in \{x, y, z\}$. The equation of a coordinate of $C_{QBp}(t)$ is as follows:

$$C_{QBp}(t) = \begin{bmatrix} t^2 & t & 1 \end{bmatrix} \begin{bmatrix} 0.5 & -1 & 0.5 \\ -1 & 1 & 0.5 \\ 0.5 & 0 & 0 \end{bmatrix} \begin{bmatrix} P_{p1} \\ P_{p2} \\ P_{p3} \end{bmatrix} \tag{2}$$

As such, for a quadratic B-spline, $[t] = [t^2\ t\ 1]$, $[G] = [P_{p1}\ P_{p2}\ P_{p3}]^T$, and [M] is a 3 × 3 matrix, as shown in Equation (2) between [t] and [G].

If one needs to find out the tangent of $C_{QBp}(t)$, denoted as $C'_{QBp}(t)$, [t] is replaced by a matrix denoted as [t]'. The elements of [t]' are the first derivative of the elements of [t]. Thus, $C'_{QBp}(t)$ is given as follows:

$$C'_{QBp}(t) = \begin{bmatrix} 2t & 1 & 0 \end{bmatrix} \begin{bmatrix} 0.5 & -1 & 0.5 \\ -1 & 1 & 0.5 \\ 0.5 & 0 & 0 \end{bmatrix} \begin{bmatrix} P_{p1} \\ P_{p2} \\ P_{p3} \end{bmatrix} \tag{3}$$

An instructor can assign students to construct the equations of other B-spline curves using the concept of shape matrices as well as the tangent vectors. This way, the students learn more intensely how to apply the knowledge of mathematics and science to an engineering problem.

Sometimes, a single quadratic B-spline curve is not good enough to model the entire object (or any other parametric curves in this matter). In this case, the modeling is carried out by using several curves. Suppose that one models a shape using a finite set of B-spline curves that are denoted as $C_{j,QBp}(t)$, j = 1, 2, ... As such, all of the possible pairs of two consecutive curves denoted as $C_{j,QBp}(t)$ and $C_{j+1,QBp}(t)$ are connected in a piecewise manner, fulfilling the C^0 and C^1 continuity [19,20]. This means that the last point of a curve is equal to the first point of the next curve (C^0 continuity), and the slope at the last point of a curve equal to the slope at the last point of the next curve (C^1 continuity). Thus, the following proposition must be true in the modeling process:

$$(C_{j,QBp}(t = 1) = C_{j+1,QBp}(t = 0)) \wedge \left(C'_{j,QBp}(t = 1) = C'_{j+1,QBp}(t = 0)\right) \tag{4}$$

The assignments based on the formulation defined in Equation (4) help students achieve the outcomes (2) and (3), i.e., an ability to design a system, component, or process to meet the desired need, and an ability to identify, formulate, and solve engineering problems. It is worth mentioning that the students have the choices of using other parametric curves (e.g., Bezier curves) to model the given object or shape. However, basic concept remains the same, as described above.

In addition, given a 2D parametric curve, as defined by Equation (1), the students can be assigned to derive an offset curve denoted as D(t), as schematically illustrated in Figure 2a. The offset curve becomes the tool-path for machining the shape (i.e., C(t)). The CNC program ensures that the center of the cutting tool follows D(t). Thus, this information is taken as the cutter locations for constructing a CNC program (Figure 1).

Now, for determining D(t), one first needs to determine the normal vector N(t) for each set of two consecutive points on C(t). The distance between C(t) and D(t) is equal to the radius of the cutting tool that is denoted as R. Thus, the expression of the offset curve D(t) is as follows:

$$D(t) = C(t) + R \frac{N(t)}{\|N(t)\|} \tag{5}$$

Figure 2b schematically illustrates how one can calculate a point denoted as $D(t1) = (D_x(t1), D_y(t1)))$ on the offset curve D(t) using two points denoted as $C(t1) = (C_x(t1), C_y(t1))$ and $C(t2) = (C_x(t2), C_y(t2))$ so that $t1, t2 \in [0,1]$ and $t2 = t1 + \Delta t$, where Δt is a very small positive number (e.g., 0.01).

Since the line passing through the points $D(t1)$ and $C(t1)$ and the line passing through the points $C(t1)$ and $C(t2)$ are orthogonal to each other, the coordinates of $D(t1)$ can be calculated as follows:

$$D_x(t1) = C_x(t1) \pm R\frac{C_y(t2) - C_y(t1)}{d(t1)} \qquad D_y(t1) = C_y(t1) \mp R\frac{C_x(t2) - C_x(t1)}{d(t1)} \tag{6}$$

In Equation (6), $d(t1)$ denotes the distance between the points $C(t1)$ and $C(t2)$, which is calculated as follows:

$$d(t1) = \sqrt{(C_x(t2) - C_x(t1))^2 + (C_y(t2) - C_y(t1))^2} \tag{7}$$

It is worth mentioning that the students need to decide the combinations of the signs (plus–minus or minus–plus) to use the Equation (6). The combination depends on the orientations of the points denoted as $C(t1)$ and $C(t2)$ (i.e., clockwise or anticlockwise direction). The case shown in Figure 2b results in a clockwise direction. For this case, the combination of signs is minus and plus for calculating $D_x(t1)$ and $D_y(t1)$, respectively. At the same time, the student may recall the vector algebra while validating Equation (6), particularly the concepts of unit vector, vector dot product, direction cosine, and alike.

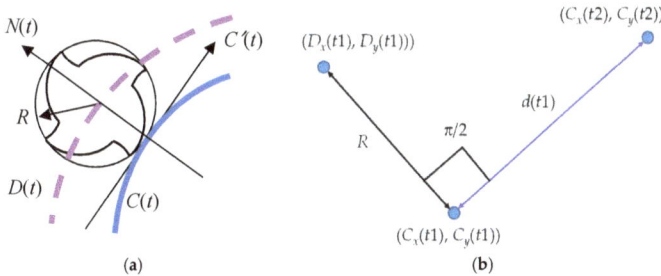

(a) (b)

Figure 2. The concept of two-dimensional (2D) tool offsetting. (**a**) The concept of offset; (**b**) calculating a point on the offset curve.

As an exercise, the students can consider the airfoil shape shown in Figure 3. The instructor can introduce the shape to the students by plotting it on a graph paper, as shown in Figure 3. Later, the students can manufacture the part, as shown in Figure 4. In this case, the students first create the CNC program and use it for machining the part using a CNC machine tool (Figure 1). The students can be given two options to do this, as described below.

In the first option, the students reconstruct the shape shown in Figure 3 using a CAD software and input the CAD data to a CAD/CAM software for generating the CNC program. In this case, the performance of the students can be determined by an output, as shown in Figure 5. This is the screenprint of the tool-path generated by commercially available software (MasterCAM™) from the CAD data of the shape that is shown in Figure 3. This type of learning ensures that the students can use the techniques, skills, and modern engineering tools that are necessary for engineering practice. However, the students' ability to apply the mathematical and engineering science knowledge gained in other courses cannot be tested or enhanced by using such an exercise. Thus, the students should be given the other option to create the tool-path by applying Equations (1)–(7). The emphasis should be given to both how to model the shapes using some piecewise B-spline curves, and then to create the offset curve $D(t)$. The students should create the CNC program, too, using the information of offset curve $D(t)$. The screenprint shown in Figure 6 shows one of the expected achievements of the students. The CNC program can be observed in Figure 6. Comparing the results obtained from the commercial CAD/CAM package (Figure 5) and the spreadsheet-based analytical approach (Figure 6), the instructor can evaluate the performance of the students and explain other issues (e.g., how the tool should approach the blank for the first time, how the tool should exit once the countering is done, how to optimize the cutting conditions, how to select the optimal cutting tool, and alike).

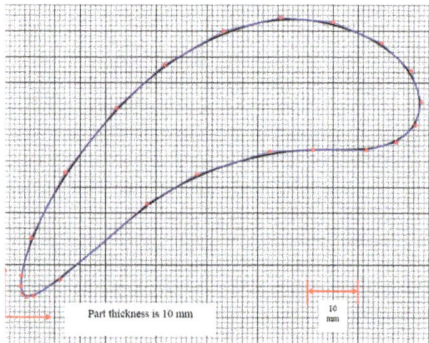

Figure 3. The schematic diagram of the shape of an airfoil.

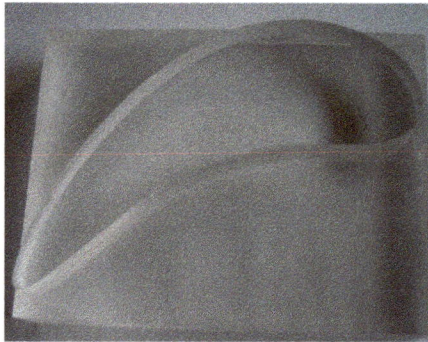

Figure 4. The shape of airfoil manufactured by Computer Numerically Controlled (CNC) machining.

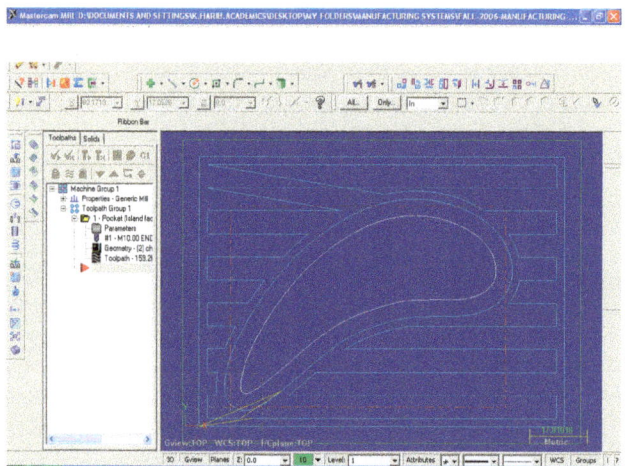

Figure 5. Tool-path generation by using a commercially available CAM software.

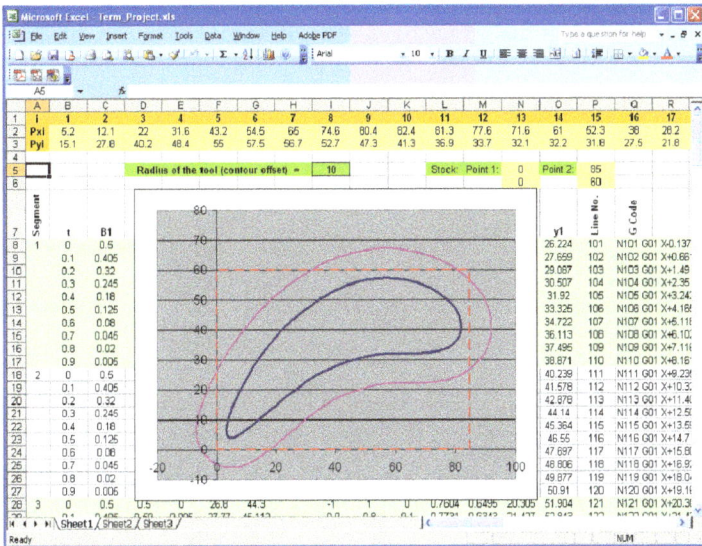

Figure 6. Tool radius offsetting using CAD/CAM package and spreadsheet application.

3. Tutorial for 3D Tool-Path Generation

This section describes a tutorial for 3D tool-path generation. This is an advanced topic compared to the previous one. The 3D tool-paths are an integral part of modern manufacturing, because freeform parametric surfaces are extensively used to model a large variety of shapes [19,20]. For constructing a freeform surface, a surface patch is generated from the product of two sets of parametric curves for two parameters, u and v. Therefore, the equation of a freeform parametric surface denoted as $C(u,v)$ is as follows:

$$C(u, v) = \sum_{i=1}^{n} \sum_{k=1}^{m} P_{ik} \, B_{ik}(u) B_{ik}(v) \quad u, v \in [0, 1] \tag{8}$$

In Equation (8), P_{ik} are the control points or vertices, and $B_{ik}(u)$ and $B_{ik}(v)$ are the blending functions. The expression of $C(u,v)$ in matrix form is as follows:

$$C(u, v) = [u][M][P][M]^T[v] \tag{9}$$

In Equation (9), $[u]$ is a row vector having n elements, $[M]$ is an $n \times m$ shape vector, $[P]$ is an $n \times m$ vector of one of the coordinates of the control points, and $[v]$ is a column vector having m elements.

For a cubic Bezier surface [19,20], the freeform surface denoted as $C_{CB}(u, v)$ is given by the following equation:

$$C_{CB}(u, v) = \begin{bmatrix} u^3 & u^2 & u & 1 \end{bmatrix} [M]_{CB} [P]_{CB} [M]_{CB}^T \begin{bmatrix} v^3 & v^2 & v & 1 \end{bmatrix}^T \tag{10}$$

In Equation (10), the shape matrix $[M]_{CB}$ is given as follows:

$$[M]_{CB} = \begin{bmatrix} -1 & 3 & -3 & 1 \\ 3 & -6 & 3 & 0 \\ -3 & 3 & 0 & 0 \\ 1 & 0 & 0 & 0 \end{bmatrix} \tag{11}$$

Now, the question is: how to create the tool-path to machine a surface similar to the one defined in Equation (10)? The machining of the surfaces similar to the one defined in Equation (10) is often done by using a ball-nose end-mill cutter, as schematically illustrated in Figure 7. The center of the ball-nose moves along an offset surface denoted as $D(u,v)$ (Figure 7). As shown in Figure 7, generating the offset surface $D(u,v)$ means that each point on $C(u,v)$ is mapped into its corresponding point on $D(u,v)$ by shifting it along the normal direction of $N(u,v)$. This requires two tangential vectors that intersect at the point, as shown in Figure 7. Thus, the expressions of the tangent vectors in the u and v directions, respectively, are as follows:

$$C_u'(u,v) = \frac{\partial C(u,v)}{\partial u} \tag{12}$$

$$C_v'(u,v) = \frac{\partial C(u,v)}{\partial v} \tag{13}$$

The normal vector $N(u,v)$ is thus obtained from the cross-product of the two tangential vectors resulting in the following expression:

$$N(u,v) = C_v'(u,v) \times C_u'(u,v) \tag{14}$$

Once the normal direction is determined, the offset surface is determined similarly to the previous case, which is as follows:

$$D(u,v) = C(u,v) + R\,\frac{N(u,v)}{\|N(u,v)\|} \tag{15}$$

In Equation (15), R is the radius of the ball-nose of the end-mill cutter.

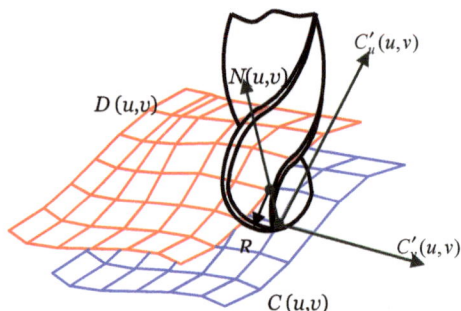

Figure 7. Tool offsetting for machining a freeform surface using a ball-nose end-mill cutter.

To exercise the above-mentioned offsetting technique for freeform surfaces, the students can use a programming tool to generate the codes required for the CNC program. Appendix A shows the required program written in MATLAB^{TM} in terms of two functions. The function shown in Figure A1 creates the offset surface from the given Bezier surface in accordance with Equations (10)–(15). It also plots the Bezier and offset surfaces. On the other hand, the function shown in Figure A2 is used for generating the codes (G-codes) for the CNC programming. It writes the code to a text file, which can be edited based on the requirement of the given CNC controller. Needless to say, the function shown in Figure A2 takes the necessary outputs from the function shown in Figure A1.

Figures 8 and 9 show the expected outcomes of the tutorial. In particular, Figure 8 shows the given and offset surfaces, which is the output of the function shown in Figure A1. For this case, the coordinates of the control points are shown below in the form of matrices.

$$[P_x] = \begin{bmatrix} 0 & 20 & 40 & 60 \\ 0 & 20 & 40 & 60 \\ 0 & 20 & 40 & 60 \\ 0 & 20 & 40 & 60 \end{bmatrix} \text{mm}, \quad [P_y] = \begin{bmatrix} 0 & 0 & 0 & 0 \\ 20 & 20 & 20 & 20 \\ 40 & 40 & 40 & 40 \\ 60 & 60 & 60 & 60 \end{bmatrix} \text{mm}, \quad [P_z] = \begin{bmatrix} 0 & -15 & -15 & 0 \\ 0 & 0 & 0 & 0 \\ -10 & 0 & 0 & -10 \\ -10 & 10 & -10 & 0 \end{bmatrix} \text{mm}$$

Figure 9 shows the freeform surface manufactured by using a three-axis CNC machine tool (Figure 1). The sharp marks seen on the manufactured freeform surface are the tool-path marks. This is due to the coarseness of the increment of u and v (the increment was 0.05) and tool diameter (offsetting for a 4-mm radius of the ball-nose end-mill cutter). The students can be assigned to make necessary changes in the formulation to reduce the sharpness of the marks. Similar to the previous tutorial, the students can achieve all of the outcomes (the outcomes denoted as (1)–(4)) mentioned above by solving the problems underlying this assignment.

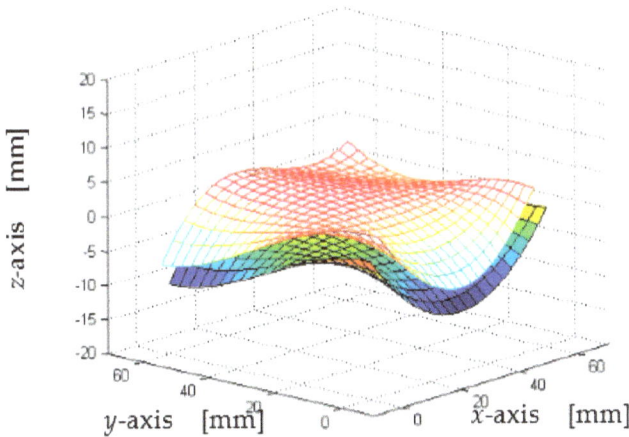

Figure 8. The given freeform surface it its offset.

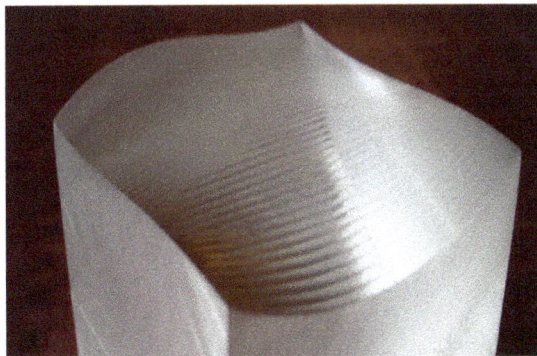

Figure 9. Manufactured freeform surface using CNC machining.

4. Programming Advanced Machine Tools

The previous two tutorials are meant for educating students with the knowledge and skills of CAD/CAM technology using conventional machine tools with a serial kinematics structure. This is not the end. The field is evolving very fast. From this perspective, the advances in the CNC machine tools can be considered. In particular, as a replacement of the conventional CNC machine tools (having a serial kinematics structure), researchers have been studying the CNC machine tools that have parallel and hybrid kinematics structures [21,22]. These machine tools are more economical, light, and functional compared to the conventional ones. Moreover, in the near future, these machine tools are expected to play their role in manufacturing.

The previous two tutorials are meant for educating students with the knowledge and skills of CAD/CAM technology that can be implemented using the conventional machine tools. Note that the conventional machine tools have a serial kinematics structure. These machine tools are by nature heavy and less economical. However, researchers have been studying a new type of CNC machine tools that have parallel or hybrid kinematics structures [21,22]. These machine tools are more economical, light (portable), and functional compared to the conventional ones, and thereby are expected to play their roles in the near future. Therefore, the students majoring in manufacturing, mechatronics, and robotics are likely to engage in scholarly activities where the parallel and hybrid kinematics' structure-driven machine tools are conceptualized, analyzed, designed, and implemented. As a result, the relevant senior-level undergraduate and graduate-level courses must prepare the students with the knowledge and skills of parallel and hybrid kinematics structures. Based on this contemplation, this section presents a tutorial that is meant for educating interested students with the knowledge of how to program a parallel kinematics machine tools for machining a 3D shape. The description of the tutorial is as follows:

Consider CNC machine tools [21] that have parallel kinematics structures, as shown in Figure 10. As seen in Figure 10, the cutting motion is controlled by six extendable linear actuators that carry the cutting tool platform. See Harib et al. [21] for more detail on the kinematics analysis and the trajectory planning. However, for the *i*-th actuator, the extension l_i is computed as follows:

$$l_i = \left\| \begin{bmatrix} X \\ Y \\ Z \end{bmatrix} + \begin{bmatrix} c^2A + cBs^2A & sAcA - cBsAcA & sBsA \\ cAsA - cBcAsA & s^2A + cBc^2A & -sBcA \\ -sAsB & cAsB & cB \end{bmatrix} \begin{bmatrix} x_{ai} \\ y_{ai} \\ z_{ai} \end{bmatrix} - \begin{bmatrix} X_{bi} \\ Y_{bi} \\ Z_{bi} \end{bmatrix} \right\| \quad (16)$$

In Equation (16), *s* and *c* denote the sine and cosine functions, respectively; $(X_{bi}, Y_{bi}, Z_{bi})^T$ is the constant position vector of the *i*-th attachment point b_i with respect to the world coordinate frame *W*; $(x_{ai}, y_{ai}, z_{ai})^T$ is the constant position vector of the *i*-th attachment point a_i with respect to the cutting tool coordinate frame *C*; $(X, Y, Z)^T$ is the position vector of the tip of the cutting tool with respect to *W*; and *A* and *B* are the angles that define the orientation of the cutting tool with respect to *W*; *A* is the horizontal rotation of the vertical plane that contains the cutting tool axis, and *B* is the tilting angle of the cutting tool inside that plane.

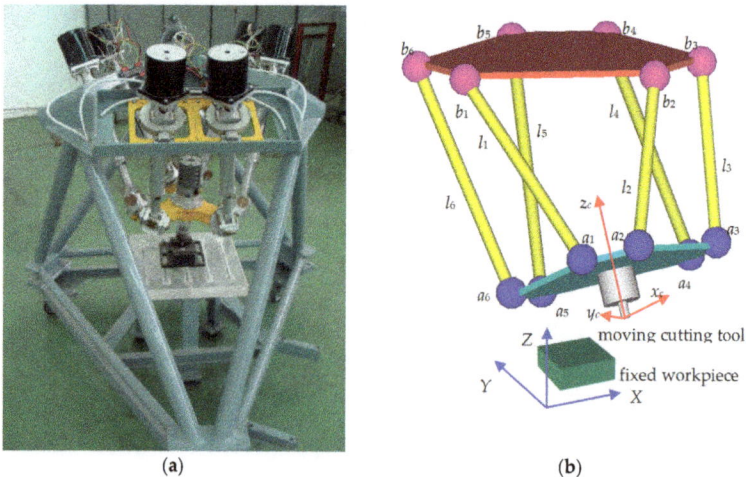

Figure 10. A parallel kinematics-driven machine tool structure [21]. (**a**) prototype; (**b**) operational parameter.

To machine a freeform surface similar to the one described in the previous section, the following procedure can be used.

Step 1: Calculate *X*, *Y*, and *Z* from the three coordinates of the parametric surface *C*(*u*,*v*).

Step 2: Calculate the normal vector $N(u,v)$ according to Equation (14).
Step 3: Calculate the unit vector along the cutting tool $n(u,v)$ as follows:

$$n(u,v) = \frac{N(u,v)}{\|N(u,v)\|} \tag{17}$$

The third column of the rotation matrix in Equation (16) is the unit vector along the axis of the cutting tool. This means that:

$$n(u,v) = \begin{bmatrix} n_x & n_y & n_z \end{bmatrix}^T = \begin{bmatrix} sBsA & -sBcA & cB \end{bmatrix}^T \tag{18}$$

Step 4: Determine A and B, as follows:

$$A = \tan^{-1}\left(\frac{n_x}{\sqrt{n_x^2 + n_y^2}}, \frac{-n_u}{\sqrt{n_x^2 + n_y^2}} \right) \tag{29}$$

$$B = \tan^{-1}\left(\sqrt{n_x^2 + n_y^2}, n_z \right) \tag{20}$$

It is worth mentioning that if the students use a package of five-axis milling, they will be able to determine (X,Y,Z,A,B) without any effort provided that a flat-end mill is used in the programming. However, if the students follow the above steps, they realize the whole process without any support of a commercial package. This way, the students having more advanced analytical abilities in CAD/CAM can be nurtured.

5. Concluding Remarks

The authors have been offering CAD/CAM and related courses for a long time for both undergraduate and graduate students of engineering degree programs. The authors have been observing that if the commercial packages are used for creating the tool-paths automatically, the students can gain the application ability, but they fail to gain the analytical ability, i.e., they become doers, but not knowers. On the contrary, the students can achieve the same using the presented tutorials where they do not need to use any dedicated CAD/CAM packages. This means that the presented tutorials contribute positively to the objective of strengthening students' ability to apply the knowledge of mathematics and engineering science, making students both doers and knowers together. Students also learn that designing a shape using CAD software is not everything. If the design is not converged to a CNC program, the materialization of the shape does not occur. This way, the students become aware of the activities of a product life cycle more intensely, which should be the case from the perspective of their future employers. In the presented tutorials, the students are introduced to the concept of parametric curves and surfaces. This improves their knowledge of dealing with the shapes having complex geometry. In addition, the students are exposed to hexapod-based machine tools. This makes the students aware of the CNC technology advancing rapidly. Therefore, it is essential for the students to understand which module of a system of CAM (or CAD) affects the operations of such machine tools from the perspective of the structure of a machine tool, as well as from the perspective of the shapes of the parts to be manufactured. Moreover, in the presented tutorials, the students get the opportunity to use standard computing and visualizing tools (e.g., Excel™ and MATLAB™). This makes them competent enough while using these tools in other courses, too. However, the instructors or the students have the freedom to use any computing platform and programming languages they prefer, although the descriptions here show the usages of spreadsheet and MATLAB™ coding. Having said that, it is not meant that the mathematical formulations presented in this article should not be obeyed. The first two tutorials can be offered to the senior-level undergraduate students enrolled in mechanical, manufacturing, and industrial engineering degree programs, whereas the other tutorial is recommended for the students who want to design mechatronics devices. While offering the tutorials, the authors recommend the approach called thematic integration or problem-based learning (PBL). In the case of thematic integration, the instructor may offer the tutorial in line with the contents of other courses (e.g., engineering design or mechanics course), so that the geometric shapes used in the other relevant course can be used in the proposed tutorials. While offering the tutorials as PBL, the students should be given a chance to form a team. The team then sets the objectives and outcomes based on the contents of the tutorials. Finally, the students present their teamwork for the evaluation.

Nevertheless, the presented tutorials can be modified by incorporating more outcomes (e.g., contemporary issues, teamwork, ethics, and alike) and the contents that are needed for mastering the digital manufacturing technology. Particularly, the issue of functioning effectively in a multidisciplinary team, and being aware of the professional and ethical responsibility, can also easily be accommodated more explicitly with the presented tutorials if the relevant degree program needs it. Similarly, the tutorials can easily be modified to accommodate the issues of sustainability and sustainable manufacturing [23], reverse engineering [24], design for additive manufacturing of complex shapes [25], and e-making [26]. At the same time, the issues of meaningful learning using a set of concept maps [27] can be considered. It is worth mentioning that concept maps [27] can represent the contents of the tutorials described in the article, and the students can even access them using their cell phones [4]. Since the concepts underlying any topic of manufacturing are ultimately linked to the universal concept maps, namely, the sustainability universe, the control universe, the material universe, the precision universe, and the process universe [4], one must be careful about this linkage while representing the tutorials using a set of concept maps. Otherwise, the desired learning objectives may not be materialized.

Author Contributions: Both authors, A.S.U. and K.H.H., equally contributed in all sections of this article.

Funding: This research received no external funding.

Conflicts of Interest: The authors declare no conflict of interest.

Appendix A

This Appendix shows the MATLAB™ codes needed for creating the offset surface from a cubic Bezier surface as defined in Section 3 and the relevant CNC programming. In particular, Figure A1 shows the MATLAB™ codes for tool offsetting of a Bezier freeform surface, whereas Figure A2 shows the MATLAB™ codes for CNC programming for machining the Bezier freeform surface.

```
function [Cx,Cy,Cz,Dx,Dy,Dz] = ...          ppvx = Vp*M*Px*M'*U';
        surface(Px,Py,Pz,offset,step_size)  ppvy = Vp*M*Py*M'*U';
M = [-1 3 -3 1;3 -6 3 0;-3 3 0 0;1 0 0 0];  ppvz = Vp*M*Pz*M'*U';
i = 0;                                      ppux = V*M*Px*M'*Up';
for v = 0:step_size:1                        ppuy = V*M*Py*M'*Up';
    i = i + 1;                               ppuz = V*M*Pz*M'*Up';
    j = 0;                                   N = cross([ppux;ppuy;ppuz], ...
    V = [v^3, v^2 ,v, 1];                    [ppvx;ppvy;ppvz]);
    Vp = [3*v^2, 2*v, 1, 0];                 PO = P + (N/norm(N)) * offset;
    for u = 0:step_size:1                    Dx(i,j) = round(1000*PO(1))/1000;
        j = j + 1;                           Dy(i,j) = round(1000*PO(2))/1000;
        U = [u^3, u^2, u, 1];                Dz(i,j) = round(1000*PO(3))/1000;
        Up = [3*u^2, 2*u, 1, 0];         end
        px = V*M*Px*M'*U';           end
        py = V*M*Py*M'*U';           mesh(Dx,Dy,Dzz)
        pz = V*M*Pz*M'*U';           hold
        P=[px;py;pz];                surf(Cx,Cy,Cz)
        Cx(i,j) = px;                hold
        Cy(i,j) = py;
        Cz(i,j) = pz;
```

Figure A1. The MATLAB^{TM} codes for tool offsetting for a Bezier freeform surface.

```
function [G_code] = ...                              for j = 2:n
Gcode(Dx,Dy,Dz, offset, start_line_num,                  count = count + 1;
file_name)                                               counter = num2str(count);
G_code = char(...                                        k = n + 1 - j;
'Manually write the initial part of the G-               if sign > 0, k = j; end
code');                                                  x_value = num2str(Dx(i,k));
[m,n]=size(Dx);                                          y_value = num2str(Dy(i,k));
count = start_line_num;                                  z_value = num2str(Dz(i,k) - offset);
for i=1:m                                                if Dx(i,k)>=0,   x_value = strcat('+',x_value);
    count = count + 1;                                   end
    counter = num2str(count);                            if Dy(i,k)>=0,   y_value = strcat('+',y_value);
    sign = (-1)^(i+1);                                   end
    k = n;                                               if (Dz(i,k)-offset)>=0, z_value = ...
    if sign > 0, k = 1; end                                  strcat('+',z_value); end
    x_value = num2str(Dx(i,k));                          current_line = strcat('N', counter,...
    y_value = num2str(Dy(i,k));                              ' G01 X',x_value,' Y',y_value,' Z',z_value);
    z_value = num2str(Dz(i,k) - offset);                 G_code = char(G_code, current_line);
    if Dx(i,k)>=0, x_value = strcat('+',x_value);    end
    end                                              end
    if Dy(i,k)>=0,                                   G_code = char(G_code, ...
        y_value = strcat('+',y_value);   end             'manually write the last part of the code');
    if (Dz(i,k)-offset)>=0,                          fid=fopen(file_name, 'w');
        z_value = strcat('+',z_value); end           for i=1:length(G_code)
    current_line = strcat('N', counter, ...              fprintf(fid,'%s\r\n',G_code(i,:));
    ' G01 X',x_value,' Y',y_value, ...               end
    ' Z',z_value);                                   fclose(fid);
    G_code = char(G_code, current_line);
```

Figure A2. The MATLAB™ codes for generating the CNC program for machining a Bezier freeform surface.

References

1. Melkanoff, M.A.; Puhl, F.; Langer, V.; Greenberg, D.P.; Shepard, M.S.; Voelcker, H. The challenge of CAD/CAM education. *SIGGRAPH Comput. Graph.* **1982**, *16*, 209–211. [CrossRef]
2. Lamancusa, J.S.; Zayas, J.L.; Soyster, A.L.; Morell, L.; Jorgensen, J. 2006 Bernard M. Gordon Prize Lecture*: The Learning Factory: Industry-Partnered Active Learning. *J. Eng. Educ.* **2008**, *97*, 5–11. [CrossRef]
3. Abele, E.; Chryssolouris, G.; Sihn, W.; Metternich, J.; ElMaraghy, H.; Seliger, G.; Sivard, G.; ElMaraghy, W.; Hummel, V.; Tisch, M.; et al. Learning factories for future oriented research and education in manufacturing. *CIRP Ann.* **2017**, *66*, 803–826. [CrossRef]
4. Ullah, A.M.M.S. On the interplay of manufacturing engineering education and e-learning. *Int. J. Mech. Eng. Educ.* **2016**, *44*, 233–254. [CrossRef]
5. Gómez-Galán, M.; Carreño-Ortega, Á.; López-Martínez, J.; Callejón-Ferre, Á.-J. Solving Power Balance Problems in Single-Traction Tractors Using PTractor Plus 1.1, a Possible Learning Aid for Students of Agricultural Engineering. *Educ. Sci.* **2018**, *8*, 68.
6. García, R.R.; Quirós, J.S.; Santos, R.G.; Peñín, P.I.Á. Teaching CAD at the university: Specifically written or commercial software? *Comput. Educ.* **2007**, *49*, 763–780. [CrossRef]

7. Tan, P.; Wu, H.; Li, P.; Xu, H. Teaching Management System with Applications of RFID and IoT Technology. *Educ. Sci.* **2018**, *8*, 26. [CrossRef]
8. Araki, T. Basic Education of CAD/CAM Through Multimedia and Network Aid. *J. Geom. Graph.* **1999**, *2*, 055–064.
9. Zhao, J.; Zhang, H.; Ma, L.; Chen, J.; Wen, Q.; Yin, S.; Wu, X. Basic CAD/CAM Engineering Education through Multimedia and Network Aid. In *Frontiers in Computer Education*; Sambath, S., Zhu, E., Eds.; Springer: Berlin/Heidelberg, Germany, 2012; pp. 33–37.
10. De Weck, O.L.; Kim, I.Y.; Graff, C.; Nadir, W.; Bell, A. A Rewarding CAD/CAE/CAM and CDIO Experience for Undergraduates. In Proceedings of the First Annual Conceive-Design-Implement-Operate (CDIO) Conference, Kingston, ON, Canada, 7–8 June 2005.
11. Ishitsuka, K.; Arai, M.; Chida, K.; Koshimizu, M. 9-102 Practice of CAD/CAM Education using 3D-CAD Systems: Second Report: Development of Teaching-Materials for NC Machining by CAM Function. *JSEE Annu. Conf.* **2005**, *53*, 132–133.
12. Lin, T.; Ullah, A.M.M.S.; Harib, K.H. On the effective teaching of CAD/CAM at the undergraduate level. *Comput.-Aided Des. Appl.* **2006**, *3*, 331–339. [CrossRef]
13. Alemzadeh, K. A Team-Based CAM Project Utilising the Latest CAD/CAM and Web-Based Technologies in the Concurrent Engineering Environment. *Int. J. Mech. Eng. Educ.* **2006**, *34*, 48–70. [CrossRef]
14. Kaneko, T.; Kashimoto, H. Development of Assistant Teaching Materials in CAD/CAM Education (Part II): Preparing CAM learning environments by LinuxCNC. *Gunma-Kohsen Rev.* **2012**, *31*, 5–10.
15. Abouelala, M.; Janan, M.T.; Brandt-Pomares, P. Methodology of selecting CAM software package for education based on a questionnaire. *Int. J. Mech. Eng. Educ.* **2015**, *43*, 77–93. [CrossRef]
16. Moschonissiotis, S.; Vosniakos, G.-C. Analysis and structuring of process planning logic for CNC machining: An education and training perspective. *Int. J. Mech. Eng. Educ.* **2018**, *46*, 97–117. [CrossRef]
17. Main, D.; Staines, H.J. An empirical approach to the assessment of cadcam. *Comput. Educ.* **1995**, *25*, 53–57. [CrossRef]
18. Hamade, R.F.; Artail, H.A.; Jaber, M.Y. Evaluating the learning process of mechanical CAD students. *Comput. Educ.* **2007**, *49*, 640–661. [CrossRef]
19. Anand, V.B. *Computer Graphics and Geometric Modeling for Engineers*; Wiley: New York, NY, USA, 1993.
20. Piegl, L.; Tiller, W. *The NURBS Book*; Springer: Berlin/Heidelberg, Germany; New York, NY, USA, 1997.
21. Harib, K.H.; Sharif Ullah, A.M.; Hammami, A. A hexapod-based machine tool with hybrid structure: Kinematic analysis and trajectory planning. *Int. J. Mach. Tools Manuf.* **2007**, *47*, 1426–1432. [CrossRef]
22. Harib, K.H.; Kamal, A.F.M.; Ullah, A.M.M.S.; Salah, Z. Parallel, Serial and Hybrid Machine Tools and Robotics Structures: Comparative Study on Optimum Kinematic Designs. In *Serial and Parallel Robot Manipulators—Kinematics, Dynamics, Control and Optimization*; Kucuk, S., Ed.; INTECH Open Access Publisher: London, UK, 2012; pp. 109–124.
23. Ullah, A.M.M.S.; Takeshi, A.; Masahiro, F.; Chowdhury, M.A.K.; Akihiko, K. Strategies for Developing Milling Tools from the Viewpoint of Sustainable Manufacturing. *Int. J. Autom. Technol.* **2016**, *10*, 727–736. [CrossRef]
24. Ullah, A.M.M.; Watanabe, M.; Kubo, A. Analytical Point-Cloud Based Geometric Modeling for Additive Manufacturing and Its Application to Cultural Heritage Preservation. *Appl. Sci.* **2018**, *8*, 656.
25. Sharif Ullah, A.M.M. Design for additive manufacturing of porous structures using stochastic point-cloud: A pragmatic approach. *Comput.-Aided Des. Appl.* **2018**, *15*, 138–146. [CrossRef]
26. Loy, J. eLearning and eMaking: 3D printing blurring the digital and the physical. *Educ. Sci.* **2014**, *4*, 108–121. [CrossRef]
27. Novak, J.D. *Learning, Creating, and Using Knowledge: Concept Maps as Facilitative Tools in Schools and Corporations*; Routledge: New York, NY, USA, 2012.

© 2018 by the authors. Licensee MDPI, Basel, Switzerland. This article is an open access article distributed under the terms and conditions of the Creative Commons Attribution (CC BY) license (http://creativecommons.org/licenses/by/4.0/).

education sciences

MDPI

Concept Paper

An Extended Constructive Alignment Model in Teaching Electromagnetism to Engineering Undergraduates

Ashanthi Maxworth[ORCID]

Physics and Engineering Physics, University of Saskatchewan, Saskatoon, SK S7N5E2, Canada; asm468@mail.usask.ca; Tel.: +1-306-966-6445

Received: 30 May 2019; Accepted: 22 July 2019; Published: 25 July 2019

check for updates

Abstract: Introducing abstract concepts to students from applied fields can be challenging. Electromagnetics is one of those courses where abstract concepts are introduced. This work presents a conceptual model which defines learning objectives in three levels for Engineering Electromagnetics. Each level is aligned with its own assessment and evaluation methods. The advantage is that the three-level learning objectives can be extended as student self-assess and instructor assessment rubrics, and a detailed implementation is presented here. This model gives students more accessibility to the learning objectives and increases the transparency of the learning and grading processes. The main goal of this conceptual model is to make students learn with the end in mind.

Keywords: engineering education; engineering classroom practices; learning objectives; student-centered learning; assessments and evaluations; rubrics; constructive alignment

---

1. Introduction

Electromagnetism is one of the critical fields in physics. It discusses the inter-dependence of electric and magnetic fields and how they behave under certain conditions [1]. Electromagnetism spans across a wide range including but not limited to communication systems (wireless, satellite, global navigation), space weather monitoring, and bio-medical applications such as body area networks, glucose monitoring using micro-strip patch antennas, hyper/hypo thermic applications, and the list continues. In today's world where everything is wireless, knowing the fundamentals of electromagnetics is essential for both physics and electrical engineering students. Based on the major the depth covered may differ. Teaching electromagnetic theory for engineering students is challenging since engineering students like to apply the knowledge to real-world scenarios, but the subject itself contains abstract mathematics. Therefore, it is a daunting task for the teachers to find the right balance between the theory and examples, given the limited time.

Electromagnetism, especially at its initial stages, requires visualization. Since most of the concepts such as "an electric charge in free space" cannot be seen, students have to imagine it in their mind. This issue together with the complexity of mathematics make students demotivated at the beginning of the semester. To avoid that issue, students need to be involved in the teaching and learning process since the beginning of the course. The more abstract a course can be the more student engagement is needed to keep the interest and the intensity of the subject [1–3]. Students should know what is expected from them and the process of achieving those. The assessment and evaluation criteria should be available to them to make the grading process transparent. This is where a constrictive alignment approach can be used.

Constructive alignment is an approach used to match the evaluation methods with the learning objectives [4–11]. In constructive alignment, students are aware of the expectations at the beginning of the semester. This process makes the teaching and learning process transparent to the students. Students know that each assessment or evaluation method is aligned with a course objective. Many universities have applied this method successfully to their courses [1,10,12–14]. Learning objectives, assessment, and evaluation methods are essential in the accreditation process. Nevertheless, average undergraduate students have less knowledge on learning objectives. The efficiency of the learning process will increase to a great extent if the students raise the concern about the learning objectives and use those to assess them.

In this work, the author presents a model where students can self-assess whether they have met the learning objectives. This model differs from the existing ones since, instead of defining a single set of learning objectives, those are given in three levels for each chapter covered. The author also suggests proper assessment and evaluation methods for each level of learning objectives. The advantage of this three-level constructive alignment model is the learning objectives can be extended to be the assessment rubrics for both students and the teacher. This approach makes the learning objectives more visible to the students and the learning process more transparent. Although this paper uses engineering electromagnetics as an example, this model can be used in teaching any major course (for example, electromagnetics) to students from an applied field (for example, electrical engineering).

2. Proposed Model

The extended constructive alignment model I am presenting here contains the following main points.

2.1. Entry Survey

At the beginning of the semester, the teacher is supposed to give the students and the entry survey. Appendix A shows a sample questionnaire prepared by the author for this paper This step builds the rapport between the teacher and the students and provides a rough understanding of the group of learners.

Knowing the plans of the students helps decide the extent of the material needing to be covered, assign reading and homework problems, and extra homework.

By identifying the preferred learning style, the teacher can adjust the course material to cater for all students. The learning style can change from topic to topic. But each person has a dominant learning style. These are diverging (learning by watching and feeling), assimilating (learning by watching and thinking), converging (learning by doing and thinking), and accommodating (learning by doing and feeling). Although some students might categorize themselves as a hybrid, it is important to know the preferred learning style of the students. When assigning students for group projects, the teacher can assign students such that each group contains one student from each learning style. The projects usually include an equal amount of workload from each learning category, and in this case workload can be successfully distributed, and students can learn from each other creating a collaborative learning environment.

Appendix A shows an entrance survey designed for an undergraduate electromagnetic course. But a revised version can be used for any subject. It is essential to keep the language of the questionnaire friendly and straightforward since this is the first written communication between the teacher and the student. For a course such as electromagnetics, students might have already heard the toughness of the class from their peers, hence it is crucial to building the rapport between the students and the teacher by using friendly language.

The entry survey also serves the purpose of making the minds prepared to absorb the complicated material by giving them a "heads-up".

2.2. The Three-Level Learning Objectives

The main difference between this proposed model and the current outcome-based education is that the current models contain only one set of learning objectives for each lesson. Whereas in this proposed model, we are introducing three levels of learning objectives. These learning objectives must be made visible to the students at the beginning of each chapter.

The learning objectives are designed as low, medium, and high. The low-level learning objectives target the basic understanding of the course material. The wordings are selected based on Bloom's taxonomy [4–9,13], and sample learning objectives for a typical undergraduate electromagnetic course is given in Table 1.

For illustration, we have selected the topics covered in a 30 credit hours undergraduate electromagnetic course. The topics covered are:

1. Gauss's Law for electrostatic fields and Maxwell's first equation.
2. Gauss's Law for magnetostatic fields and Maxwell's second equation.
3. Faraday's Law for time-varying electric fields and Maxwell's third equation.
4. Ampere's Law for time-varying electromagnetic fields and Maxwell's fourth equation.
5. Plane wave solution.
6. Poynting theory, electromagnetic power, basic electromagnetic radiation principles, and their applications.

Depending on the university, there are variations of these topics and instructors would breakdown the above topics to chunks. But the above are considered mandatory.

The learning objectives are designed on three levels for all chapters. In a course such as engineering electromagnetics, students might have different goals. If a student is planning on going to graduate studies, he or she may want to learn more in-depth physics content whereas if someone wants to join the industry wants to know just enough material. With this method, the student can decide whether he or she has achieved the required learning objective.

Table 1. Three level learning objectives and the suggested assessment and evaluation methods (Bloom's taxonomies are in bold face).

Topic	Learning Objective(s)	Assessment Method(s)	Evaluation Method(s)
1. Gauss's Law for electrostatic fields and Maxwell's first equation.	Low: By the end of this chapter students are expected to **explain** Maxwell's first equation and its implications.	Conference, Self-assessment	A quiz Presentation, Case study
	Medium: By the end of this chapter students will be able to **apply** Maxwell's first equation to solve real-world physics problems.	Self-assessment quiz, Question and answer, I am in the fog about …	Exam problem, Quiz
	High: By the end of this chapter students are expected to **design** a basic static charge dust collector using Maxwell's first equation.	Chart it out, Concept map	Group project (2-3 students), A term paper Project report
2. Gauss's Law for magnetostatic fields and Maxwell's second equation.	Low: By the end of this chapter students will be able to **discuss** the practical implications of Maxwell's second equation.	Discussion, Conference	Short presentation, Quiz, Short essay
	Medium: By the end of this chapter students will be able to **solve** problems related to Maxwell's second equation.	Self-assessment quiz, I am in the fog about, Operation outline	Exam questions, Quizzes
	High: By the end of this chapter students will be able to **create** a computer software model of Earth's magnetic system.	Chart it out, Ticket out the door, Concept map	Problem based project, Research report, Research paper

Table 1. *Cont.*

Topic	Learning Objective(s)	Assessment Method(s)	Evaluation Method(s)
3. Faraday's Law for time-varying electric fields and Maxwell's third equation.	Low: By the end of this chapter students are expected to **describe** Faraday's law and its implications.	Discussion, Conference	Quiz Short, presentation
	Medium: Upon completing this chapter students are expected to **compute** values for real-world problems based on Faraday's law.	Self-assessment quiz, I am in the fog about, Operation outline	Exam questions, Quizzes
	High: By the end of this chapter students will be able to **build** an electromagnetic inductor to **demonstrate** Faraday's law.	Ticket out the door, Concept map	Experiment, Prototype building
4. Ampere's Law for time-varying electromagnetic fields and Maxwell's fourth equation.	Low: By the end of this chapter students are able to **define** Ampere's law and its implications.	Ticket out the door	Presentation, Short quiz answers, Short essay
	Medium: By the end of this chapter students are able to **calculate** values for a real-world application using Ampere's law.	Self-assessment quiz, I am in a fog about, Question and answer	Exam questions, Quizzes
	High: by the end of this chapter students are expected to **construct** an electromagnet with given specifications based on Ampere's law.	Conference, Ticket out the door, Concept map	Short project, Live demonstration, Presentation of a prototype
5. Plane wave solution.	Low: By the end of this chapter students are able to **state** the plane wave solution.	Ticket out the door	Presentation, Short essay, Short quiz
	Medium: By the end of this chapter students are able to **manipulate** the plane wave solution and apply it in a real-world problem.	Self-assessment quiz, Questions and answers, Operations outline	Exam questions, Quizzes
	High: By the end of this chapter students will be able to **synthesize** plane wave electromagnetic propagation in computer software.	Chart it out, Ticket out the door, Concept map	Project, Demonstration, Video presentation
6. Poynting theory, electromagnetic power, basic electromagnetic radiation principles, and their applications.	Low: By the end of this chapter students will be able to **identify** the appropriate concepts used in real-world EM wave propagation applications.	Ticket out the door, Discussion	Presentation, Essay
	Medium: By the end of this chapter students will be able to **analyze** the real-world EM applications using appropriate concepts.	Self-assessment quiz, I am in the fog about, Operations outline	Exam questions, Long answer quizzes, Summary paper
	High: By the end of this chapter students are expected to **integrate** EM concepts and **implement** a solution to a real-world problem.	Conference, I am in the fog about, Concept map	Prototype building, Video presentation, Term paper, Presentation

2.3. Three-Level Assessment and Evaluation Methods

In this paper, the author suggests assessment and evaluation methods for each level of learning objectives. The teachers can pick a process that they might think suits the class environment to add a variety and build a collaborative environment. In Table 2 below the author shows a standard grade break down. Depending on the class performance and average, the teacher is free to make necessary adjustments. This is the "judgement" component since it determines the final grade based on the performance.

Since this course is targeting towards engineering students, the highest percentage is allocated for projects. Hence the majority of the course percentage is for the application or projects. Projects can be assigned individually or as groups since there are five projects within about 14–16 weeks. Hence, appointing group projects will expedite the submissions. The statistics collected from the entry survey can be used here for assigning groups. It is preferred that each group contains at least one member from each learning style. Also, students get to know each other by changing the groups for each project.

Most of the faculty is familiar with the evaluation methods. But some of the assessment techniques are unfamiliar to most of the instructors. But assessment techniques are necessary because those give students an opportunity to determine whether they have met the learning objectives. Appendix B shows a sample self-assessment quiz for chapter 1; Appendix C is a concept map for Section 2 and Appendix D illustrates an operations outline for topic 6.

Table 2. Standard grade breakdown.

Evaluation Method	Percentage
Quizzes	10%
Homework	15%
Exams (3 including the finals)	25%
Projects (6 mini projects)	50%

2.4. Student Self-Assessment and Instructor Assessment Rubrics

Tables 3 and 4 show the student self-assessment and instructor assessment rubrics. The importance of these rubrics is that these are directly aligning with the learning objectives. Therefore, this is an extended model. With this model, students can determine where they are in the rubric, and the instructors can assess where each student is in the performance chart. The alignment between the learning objectives and the assessment rubrics extends this model and increases the visibility of the learning objectives. It also increases the transparency of the teaching-learning process.

The three levels of Bloom's taxonomy [9] (low, medium, and high) and the model introduced by Biggs in 1996 [5] are closely related. In the Biggs model the term pre-structural was used to indicate the unsatisfactory performance. The low-level learning objectives in Bloom's taxonomy are similar to the unistructural performance in Biggs' model. If the student is at this level, here the student is categorized as "needs improvement". If the student has achieved medium-level learning objectives (multi-structural in Biggs' model), he or she will be categorized into the satisfactory level. If the student has achieved the high-level learning objectives (or the relational in the Biggs model) that student will be categorized as excellent.

Table 3. Student self-assessment rubric.

Topic	Unsatisfactory	Needs Development	Satisfactory	Excellent
Gauss's Law for electrostatic fields and Maxwell's first equation.	I can neither both explain, apply nor design an application based on Maxwell's first equation.	I can **explain** Maxwell's first equation. But I can neither apply nor design an application based on it.	I can **explain** and **apply** Maxwell's first equation. But I cannot design an application based on it.	I can **explain**, **apply** and **design** an application using Maxwell's first equation.
Gauss's Law for magnetostatic fields and Maxwell's second equation.	I can neither discuss the implications, solve problems nor create an application based on Maxwell's second equation.	I can **discuss** the implications of Maxwell's second equation. But I can neither solve problems nor create an application using it.	I can **discuss** and **solve** problems using Maxwell's second equation. But I cannot create an application based on it.	I can **discuss** the implications, **solve** problems and **create** an application based on Maxwell's second equation.

Table 3. Cont.

Topic	Unsatisfactory	Needs Development	Satisfactory	Excellent
Faraday's Law for time-varying electric fields and Maxwell's third equation	I can neither describe, compute nor build an application based on Faraday's law.	I can **describe** Faraday's law. But I can neither compute nor build an application based on it.	I can **describe** and **compute** values for a practical problem. But I cannot build an application.	I can **describe, compute** and **build** an application to **demonstrate** Faraday's law.
Ampere's Law for time-varying electromagnetic fields and Maxwell's fourth equation.	I can neither, define, calculate values nor construct an application using Ampere's law.	I can **define** Ampere's law. But I cannot calculate values or construct an application.	I can **define** and **calculate** values for problems, using Ampere's law. But I cannot construct an application.	I can **define, calculate** values and **construct** an application based on Ampere's law.
Plane wave solution.	I can neither state, manipulate nor synthesize plane wave solution.	I can **state** plane wave solution. But I cannot manipulate or synthesize it.	I can **state** and **manipulate** plane wave solution. But I cannot synthesize it.	I can **state, manipulate** and **synthesize** plane wave solution.
Poynting theory, electromagnetic power, basic electromagnetic radiation principles, and their applications.	I can neither identify, analyze nor integrate practical applications of EM wave propagation concepts.	I can **identify** EM concepts for real-world scenarios. But I can neither analyze nor integrate concepts.	I can **identify** and **analyze** EM concepts for real-world scenarios. But I cannot integrate those for implementations.	I can **identify, analyze** and **integrate** EM appropriate EM concepts to **implement** solutions.

Table 4. Instructor assessment rubric.

Topic	Unsatisfactory	Needs Development	Satisfactory	Excellent
Gauss's Law for electrostatic fields and Maxwell's first equation.	Student can neither both explain, apply nor design an application based on Maxwell's first equation.	Student can **explain** Maxwell's first equation. But the student can neither apply nor design an application based on it.	Student can **explain** and **apply** Maxwell's first equation. But the student cannot design an application based on it.	Student can **explain, apply** and **design** an application using Maxwell's first equation.
Gauss's Law for magnetostatic fields and Maxwell's second equation.	Student can neither discuss the implications, solve problems nor create an application based on Maxwell's second equation.	Student can **discuss** the implications of Maxwell's second equation. But the student can neither solve problems nor create an application using it.	Student can **discuss** and **solve** problems using Maxwell's second equation. But the student cannot create an application based on it.	Student can **discuss** the implications, **solve** problems and **create** an application based on Maxwell's second equation.
Faraday's Law for time-varying electric fields and Maxwell's third equation	Student can neither describe, compute nor build an application based on Faraday's law.	Student can **describe** Faraday's law. But the student can neither compute nor build an application based on it.	Student can **describe** and **compute** values for a practical problem. But the student cannot build an application.	Student can **describe, compute** and **build** an application to **demonstrate** Faraday's law.
Ampere's Law for time-varying electromagnetic fields and Maxwell's fourth equation.	Student can neither, define, calculate values nor construct an application using Ampere's law.	Student can **define** Ampere's law. But the student cannot calculate values or construct an application.	Student can **define** and **calculate** values for problems, using Ampere's law. But the student cannot construct an application.	Student can **define, calculate** values and **construct** an application based on Ampere's law.
Plane wave solution.	Student can neither state, manipulate nor synthesize plane wave solution.	Student can **state** plane wave solution. But the student cannot manipulate or synthesize it.	Student can **state** and **manipulate** plane wave solution. But the student cannot synthesize it.	Student can **state, manipulate** and **synthesize** plane wave solution.

Table 4. *Cont.*

Topic	Unsatisfactory	Needs Development	Satisfactory	Excellent
Poynting theory, electromagnetic power, basic electromagnetic radiation principles, and their applications.	Student can neither identify, analyze nor integrate practical applications of EM wave propagation concepts.	Student can **identify** EM concepts for real-world scenarios. But the student can neither analyze nor integrate concepts.	Student can **identify** and **analyze** EM concepts for real-world scenarios. But the student cannot integrate those for implementations.	Student can **identify**, **analyze** and **integrate** EM appropriate EM concepts to **implement** solutions.

3. Discussion

This work presented an extended constructive alignment model for teaching engineering electromagnetics. The goal of this model is to relate learning objectives with the assessment rubrics, hence students become more familiar with the expectations. Because a typical undergraduate student would care about the course objectives if and only if it is related with the grade. Teaching electromagnetics is challenging for engineering students since most of the engineering students do not expect a course to be theoretical and mathematically complicated. But with electromagnetics being an essential field, those complex material needs to be introduced regardless of the challenges. Techniques such as the flipped classroom can be applied but those techniques might not have an effect given the work–life balance of students and the context of the course.

In this paper, the author presents a model which defines the learning objectives in three levels, hence it can be converted to assessment rubrics with minimal effort as illustrated later in this paper. Therefore, this model is an extended version of the existing constructive alignment concept. Initially, this might increase the workload of the instructor for course design, revisiting the learning objectives, preparation of assessments, and course material. But once the learning objectives are written the entire process will be much more apparent to the instructor as well as to the student. Instead of jumping right to the high-level learning objective, the students can self-guide themselves through each step of the learning process.

The difference between this model and the existing constructive alignment method is here course objectives are presented in three levels. Therefore, students know what is expected from them and the assessment and evaluation process become more transparent. As mentioned in this paper, this model has the advantage of directly converting the three-level course objectives into assessment rubrics.

Currently this model is structured in a way that everything is well defined and gives only little room for student's imaginative thinking. But as with every course, instructors have to fine-tune the model as it goes, keeping the core concepts intact. For example, if a student comes up with an alternative project instead of the one assigned, an instructor can compare the workload of the project and grant permission. And extra credit can be given if students perform beyond expectations. Another way to include creativity is the course objectives can be modified such that instead of the instructor assigning the projects students should come up with their own ideas. This second option might be better suited for an advanced undergraduate or a graduate level course where students have enough knowledge about the subject matter to challenge their creativity.

4. Conclusions

This paper presents an extended constructive alignment model for teaching electromagnetism to engineering undergraduates. Although the model is developed for electromagnetism, this can be applied when introducing any abstract field to undergraduates from an applied field. Currently this model is at a conceptual state and a follow-up paper will be published with the results of its implementation.

Funding: This research received no external funding.

Educ. Sci. **2019**, *9*, 199

Acknowledgments: The author would like to acknowledge Rebekah Bennetch at the University of Saskatchewan, Canada, and the teaching team of GPS 989: Philosophy and Practice of University Teaching.

Conflicts of Interest: The author declares no conflict of interest.

Appendix A

Class entrance survey

I _____ answer these questions to the best of my knowledge.

1. As of now, after completing my undergraduate degree, I am planning on:

   a. Doing graduate studies
   b. Working in the industry
   c. Doing my own thing, for example, painting, carving, sculpture

2. If I can get one thing out of this course it would be:

   a. Nothing, I registered just because it is required. I have a different subject interest.
   b. Try this subject and see whether I should pursue this for my graduate studies
   c. Learn how things work and apply it at work

3. While registering for this course

   a. I knew/heard this course it very mathematics and physics intensive
   b. Oops, I did not know that. But I can catch up quickly.
   c. Oh no, why? Engineers don't need math or physics.

4. We all learn in different ways. But if I have to choose one, that I learn quickly by

   a. Watching and feeling
   b. Watching and thinking
   c. Doing and feeling
   d. Doing and thinking

5. Knock on wood, but if my performances at exams are not satisfactory

   a. I will sue the instructor. I am exceptional, and it is always the instructor's fault.
   b. It might be a bad day. I want to write a make-up exam.
   c. I get nervous at exams. If so I will do an extra project or a presentation, whichever it takes to show my actual knowledge.

Appendix B

Self-assessment quiz: Gauss's Law or Maxwell's first equation

1. A spherical charge cloud with volume charge density ρ_v and radius a, is located at the origin of a spherical coordinate system. Determine the electric flux density and electric field intensity at a distance r such that,

   a. $r < a$
   b. $r \geq a$

2. An infinite length of a wire contains a line charge density of ρ_l. Choosing a suitable coordinate system, calculate the electric field intensity at a radial distance r from the wire.

3. Two hollow spheres are located at the origin of a spherical coordinate system. Surface of the inner sphere carries a total charge of $+Q$ and the surface of the outer sphere carries a total charge of $-Q$. The space between the spheres is filled with air.

a. Find the electric field intensity at a radial distance $r : a < r < b$
b. Calculate the potential difference between the two spheres
c. How much of a capacitance is developed between the two spheres?

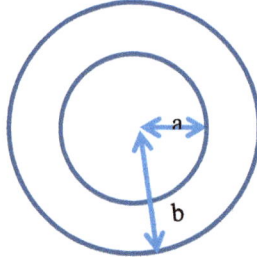

Appendix C

Concept map: Creating a Computer Software Model of Earth's Magnetic System

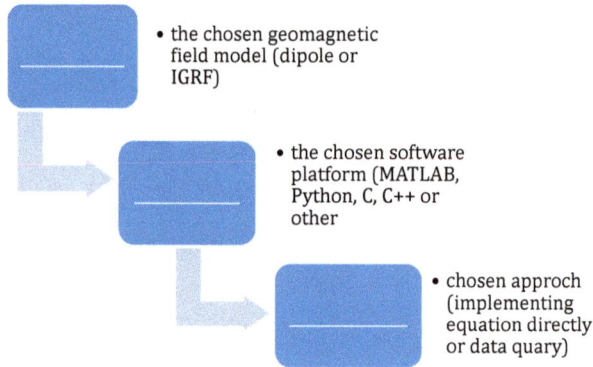

- the chosen geomagnetic field model (dipole or IGRF)
- the chosen software platform (MATLAB, Python, C, C++ or other
- chosen approch (implementing equation directly or data quary)

Appendix D

Operations Outline: Basic EM Propagation Concepts

1. A 15W EM radiator is isotopically radiating energy equally in all directions. Your task is to calculate the surface area of a dish antenna located 15 m from the radiator to collect 1W of power.

    a. What should be the radiating surface for the above radiator?

    ... ...

    b. Calculate the average power density 15 m from the radiator.

    ... ...

    c. If the goal is to collect 1W at the receiver, what should be the surface area of the receiver

    ... ...

    d. What should be the radius of the above dish

    ... ...

    e. If the above dish was replaced by a parabolic antenna with the same radius, will the power collected will increase or decrease?

    ... ...

References

1. Hadzigeorgiou, Y.; Garganourakis, V. Using Nikola Tesla's story and his experiments as presented in the film "The Prestige" to promote scientific inquiry: A report of an action research Project. *Interchange* **2010**, *41*, 363–378. [CrossRef]
2. Hadzigeorgiou, Y.; Klassen, S.; Froese-Klassen, C. Encouraging a "romantic understanding" of science: The effect of the Nikola Tesla story. *Sci. Educ.* **2012**, *21*, 1111–1138. [CrossRef]
3. Hadzigeorgiou, Y. Imaginative science education. In *The Central Role of Imagination in Science Education*; Springer International: Cham, Switzerland, 2016.
4. Airasian, P.; Cruikshank, K.A.; Mayer, R.E.; Pintrich, P.; Raths, J.; Wittrock, M.C. *A Taxonomy for Learning, Teaching, and Assessing: A Revision of Bloom's Taxonomy of Educational Objectives*; Anderson, L.W., Krathwohl, D.R., Eds.; Allyn and Bacon: Boston, MA, USA, 2001; ISBN 978-0-8013-1903-7.
5. Biggs, J.B. Enhancing teaching through constructive alignment. *High. Educ.* **1996**, *32*, 347–364. [CrossRef]
6. Biggs, J.B. What the student does Teaching for enhanced learning. *High. Educ. Res. Dev.* **1999**, *18*, 1–19. [CrossRef]
7. Biggs, J. *Aligning Teaching and Assessment to Curriculum Objectives*; Imaginative Curriculum Project, LTSN Generic Centre: York, UK, 2003.
8. Biggs, J.B. *Teaching for Quality Learning at University: What the Student Does*; McGraw-Hill: Maidenhead, UK, 2011; ISBN 9780335242757.
9. Bloom, B.S.; Engelhart, M.D.; Furst, E.J.; Hill, W.H.; Krathwohl, D.R. *Taxonomy of Educational Objectives: The Classification of Educational Goals. Handbook I: Cognitive Domain*; David McKay Company: New York, NY, USA, 1965.
10. Brooks, J.; Brooks, M. *In Search of Understanding: The Case for Constructivist Classrooms*; ASCD: Virginia, VA, USA, 1993.
11. Cain, A.; Grundy, J.; Woodward, C.J. Focusing on Learning through Constructive Alignment with Task-Oriented Portfolio Assessment. *Eur. J. Eng. Educ.* **2018**, *43*, 569–584. [CrossRef]
12. Cobb, P. Theories of knowledge and instructional design a response to Colliver. *Teach. Learn. Med.* **2002**, *14*, 52–55. [CrossRef]
13. Knaack, L. *A Practical Handbook for Educators: Designing Learning Opportunities*; De Sitter Pubns: Whitby, ON, Canada, 2011; ISBN-13: 978-1897160473.
14. Smith, C. Design–focused evaluation. *Assess. Eval. High. Educ.* **2008**, *33*, 631–645. [CrossRef]

© 2019 by the author. Licensee MDPI, Basel, Switzerland. This article is an open access article distributed under the terms and conditions of the Creative Commons Attribution (CC BY) license (http://creativecommons.org/licenses/by/4.0/).

MDPI

St. Alban-Anlage 66

4052 Basel

Switzerland

Tel. +41 61 683 77 34

Fax +41 61 302 89 18

www.mdpi.com

Education Sciences Editorial Office

E-mail: education@mdpi.com

www.mdpi.com/journal/education

www.ingramcontent.com/pod-product-compliance
Lightning Source LLC
Chambersburg PA
CBHW051859210326
41597CB00033B/5950